D1348948

"*The Blue Line Imperative* is compelling, fun and provocative. It is a must-read!"
**W. Chan Kim, bestselling author of *Blue Ocean Strategy*, The BCG Chair Professor of INSEAD and the Co-director of the INSEAD Blue Ocean Strategy Institute**

"Professors Kaiser and Young bring a unique, entertaining and irreverent approach to teaching executives what it really means to create value – including the dangers of relying on conventional wisdom about performance measurement."
**Tim Koller, author of *Valuation: Measuring and Managing the Value of Companies***

"Kaiser and Young introduce a fresh perspective on what value really means and how to manage toward it. This is an indispensable new look at the most important business issue all companies face."
**Merlin Swire, Director, John Swire & Sons, Ltd**

"In our firm, we implemented a new culture emphasizing that Each of Us Counts to get everyone motivated to contribute. When we added the focus on value creation as outlined by Kaiser and Young in *The Blue Line Imperative*, it ensured those contributions were oriented in the right direction for long-term success!"
**Alberto Grua, Senior Vice President, Grünenthal Europe, US & Australia and Management Board member**

"*The Blue Line Imperative* provides valuable lessons about what value is, how to measure it, and how to create it. Many companies follow the red line, thinking it is the right thing to do. This book explains why the red line ultimately does not work and why the blue line does."
**Steve Kaplan, Neubauer Family Distinguished Service Professor of Entrepreneurship and Finance, The University of Chicago Booth School of Business**

"*The Blue Line Imperative* is insightful, impactful, humorous and unique – and the first guide that I have ever come across that gets to the core of why leaders, managers and companies fail. Embrace and apply these concepts."

**Douglas Rosefsky, Managing Director at Alvarez & Marsal and two-time winner of the Turnaround of the Year Award (Turnaround Management Association, 2003 and 2010)**

"Kaiser and Young's *The Blue Line Imperative* connects the dots in the customer–shareholder value puzzle and delivers a resonating message on the necessity of building a data-driven company. In other words, stop chasing KPIs and start focusing on value creation that lasts."

**Martin Heijnsbroek, Managing Partner, MICompany**

# THE BLUE LINE IMPERATIVE

# THE BLUE LINE IMPERATIVE

## What Managing for Value Really Means

By Kevin Kaiser and S. David Young

**JB** JOSSEY-BASS™
A Wiley Brand

© 2013 John Wiley & Sons, Ltd

Under the Jossey-Bass imprint, Jossey-Bass, 989 Market Street, San Francisco CA 94103-1741, USA www.josseybass.com

*Registered office*
John Wiley & Sons Ltd, The Atrium, Southern Gate, Chichester, West Sussex, PO19 8SQ, United Kingdom

For details of our global editorial offices, for customer services and for information about how to apply for permission to reuse the copyright material in this book please see our website at www.wiley.com.

All rights reserved. No part of this publication may be reproduced, stored in a retrieval system, or transmitted, in any form or by any means, electronic, mechanical, photocopying, recording or otherwise, except as permitted by the UK Copyright, Designs and Patents Act 1988, without the prior permission of the publisher.

Wiley publishes in a variety of print and electronic formats and by print-on-demand. Some material included with standard print versions of this book may not be included in e-books or in print-on-demand. If this book refers to media such as a CD or DVD that is not included in the version you purchased, you may download this material at http://booksupport.wiley.com. For more information about Wiley products, visit www.wiley.com.

Designations used by companies to distinguish their products are often claimed as trademarks. All brand names and product names used in this book are trade names, service marks, trademarks or registered trademarks of their respective owners. The publisher is not associated with any product or vendor mentioned in this book.

Limit of Liability/Disclaimer of Warranty: While the publisher and author have used their best efforts in preparing this book, they make no representations or warranties with respect to the accuracy or completeness of the contents of this book and specifically disclaim any implied warranties of merchantability or fitness for a particular purpose. It is sold on the understanding that the publisher is not engaged in rendering professional services and neither the publisher nor the author shall be liable for damages arising herefrom. If professional advice or other expert assistance is required, the services of a competent professional should be sought.

*Library of Congress Cataloging-in-Publication Data*
Kaiser, Kevin.
 The blue line imperative : what managing for value really means / Kevin Kaiser and S. David Young.
  1 online resource.
 Includes index.
 Description based on print version record and CIP data provided by publisher; resource not viewed.
 ISBN 978-1-118-51089-6 (ebk) – ISBN 978-1-118-51090-2 (ebk) – ISBN 978-1-118-51088-9 (hbk)
1. Management. 2. Value. 3. Strategic planning. I. Young, S. David, 1955– II. Title.
 HD31
 658.4'01–dc23

2013020502

A catalogue record for this book is available from the British Library.

ISBN 978-1-118-51088-9 (hbk)   ISBN 978-1-118-51089-6 (ebk)
ISBN 978-1-118-51091-9 (ebk)   ISBN 978-1-118-51090-2 (ebk)

Set in 11/15 pt ITC Garamond Std Book by Toppan Best-set Premedia Limited

# Table of Contents

*Kevin Kaiser. To the thousands of MBA, EMBA and executive partici-pants I have taught who have inspired me, driven me, and helped me over so many years to develop countless insights and concepts. And to the thousands more I hope to teach in the future.*

*S. David Young. To my daughter Adiva.*

# Preface

The effort to write this book began with a conversation we had years ago about the difficulty of teaching business school participants what it means to "manage for value." In our view, this topic incorporates four essential elements: (1) What is value?; (2) Why is it important?; (3) If it's so important, why aren't managers already doing it?; and (4) How can we help managers to do it? The more time and energy we spent trying to explain, the more we realized that none of these questions has a simple answer. As we developed the material to support these efforts, and as we delivered that material to an increasing number of participants, we received more and more requests for book recommendations that would summarize the increasingly broad-thinking ideas we were teaching. Some books address the finance element while others address the accounting element, but those books explain only the "silo" dimension of value and are more about tracking and modeling value than how to manage for it. What we observe, and teach, is that managing for value requires taking a more holistic approach, to consider economic (micro and macro), scientific (biology or physics) and psychological and customer-value aspects. We found we had increasing difficulty finding books to recommend which covered the "manage for value" topic. So, in answer to the repeated requests from our students, we decided to write this book.

We had both faced the difficulty of conveying the principles of value creation to our executive and MBA participants, and realized that if these current and future managers don't understand what value is, then it is extremely unlikely they will successfully manage for value in their organizations. We noticed that when we would ask a class of 40 participants to write down a definition for value, we would usually receive 40 different answers. So how were we to write a book on managing for value if no two managers understood or agreed on what value meant? We talked to different faculty at our business school to see how they defined value. Asking faculty in Finance, Marketing, Organizational Behavior, Strategy, and other areas, it turned out that business school academics were also using different definitions (and often disagreed quite strongly with the definitions offered by their colleagues). Thus, we perceived the need to establish a common definition of value in order to have any chance of helping people manage for it.

It is relatively easy to show what is **not** managing for value – for example, managing for profit is **not** equivalent to managing for long-term value. However, knowing what value is **not** doesn't really clarify what value **is**. To answer this question, we took a somewhat unconventional "crowd-sourced" approach, to try and incorporate the different perspectives we encountered. Our aim in this book is to talk about value in a way that our academic colleagues across the disciplines can be comfortable with using in their research and in the classroom, as well as one which our seasoned executive education participants can accept and see fitting with their experience. We also chose to define it using a method of backward induction as the answer to the question: What does an organization have to accomplish in order to not end up bankrupt over a century or more? This approach enabled us to avoid the opinions and differing perspectives of individuals, and to define value in an objective way. So although each person may hold a definition of value that is personal and unique to them, we also found there is a definition that is objective and common to all of us, including organizations. We explain both in this book. We also demonstrate the critical connection between the two definitions, in which one drives the other.

Blended into the ongoing development of the effort to define value was the analysis of why it is important to manage for it, and similarly, why it is so difficult to do. The question of why it is so important to create value, which appeared self-evident at the beginning, turned out to be a rather broad discussion. As we demonstrate in this book, it is the combination of a market-based system for allocating resources and a well-functioning market for capital, which is forcing organizations to align the two definitions of value if they wish to sustain their existence. This means, simply, that in order to be confident of generating the cash to sustain the organization, it is increasingly important that the organization delivers sustainable happiness to those it serves with its products and services. In this way, the creation of value becomes an imperative (rather than a choice dependent upon what the manager feels like doing on any given day) – those organizations that create and deliver value will have a future, and those that do not, will face economic ruin. Upon closer inspection, we noticed that this system is no more, nor no less, than an evolutionary force of nature. When viewed over thousands of years, rather than a year, or a decade, or even a century, the need to deliver value, as defined herein, is clearly not a question upon which humanity can easily choose to agree or disagree. Those who create value will be defined as sustainable, and they will sustain, and those who do not will soon face extinction.

Despite this "imperative," we observe many organizations destroying value – sacrificing the longer-term sustainability of the organization to achieve a short-term target. Indeed, we like to pose the question in our executive programs to those who've worked in any of these organizations: "How many of you have knowingly destroyed value in order to deliver on a target or indicator that you've been assigned to hit?" Initially we were shocked by the high percentage of hands which were raised in confirmation that they had engaged in this behavior. (Now we are accustomed to it, so we are no longer shocked.) When asked to explain how they destroyed value, the responses were remarkably similar: it was whenever they engaged resources for an objective which could have been attained using fewer resources. This behavior was found to be rampant in all organizations; from companies and banks to charities and governments. This begs the question: Why do people destroy value if they actually know what value is? And why is it important for them to create it and avoid the demise of their organization? We devote a considerable amount of space and time in this book to this question, which we now know to be far more important than we appreciated at the outset of this project.

Once we had uncovered the "what," "why," and "why not" of value, we then tackled the profoundly difficult question: How do we orient our organizations around value as an objective? What techniques and tools can we apply in order to (1) know when we are creating value (at least, within a range), (2) encourage our people to take value-enhancing decisions, and (3) build a culture across our organization that will attract and reward those who create value? Our approach to these questions again followed a similar process as those above, where we would pose hypothetical tools, frameworks, techniques and ideas, based on insights from several areas of academia (with emphasis on psychology, neuroscience, organizational behavior, and finance), to our classroom participants. With their challenges and feedback, and based on observations of their behaviors and answers to carefully designed questions and case studies, we were able to test the validity and effectiveness of alternative techniques and hypotheses.

What emerged is presented in this book as blue line management. It seeks to incorporate insights from many areas of study to enable a manager to design, and continuously adapt, a management system and process that will keep the organization and its people oriented around value creation, while remaining highly motivated and keen to show up each day in order to do it again.

We continue this process of testing our insights and ideas, and in this way the book will never be complete. But the time has come to get this out to a broader group than those people we have in our MBA and executive classrooms, so that the benefits can be spread far and wide, and the learning can be accelerated. We hope you enjoy the journey on this path to value creation, and look forward to any and all feedback, and additional insights, which we are certain you can offer to improve still further both our understanding and our ability to effectively convey the key messages related to "managing for value."

Kevin M.J. Kaiser
S. David Young

# Acknowledgments

The efforts and support of many people made this book possible. We would like to single out Kate Kirk from Cambridge Editorial Partnership and the stellar editorial work she did in the early stages of the project. Her judgment and skill were indispensable. We would also like to thank I.J. Schecter—friend, confidant, writer, and literary agent. His efforts in getting the manuscript ready for publication were consistently outstanding. Thanks are also due to the thousands of participants in our executive and MBA programs who allowed us to test the core ideas of this book on them. Their insights and active participation strengthened the focus and key arguments of the text in so many ways.

Katherine Philips-Kaiser has been a sounding board and reliable provider of constructive criticism for years, which has improved, refined and strengthened the coherence and clarity of the themes and concepts in this book immeasurably.

And finally, we would like to thank our families for their love and support and inspiration throughout this endeavor.

# Acknowledgments

# CHAPTER ONE

# WHAT IS VALUE?

*"Oh Happiness! our being's end and aim!"*
– Alexander Pope, English poet, 1688–1744,
*An Essay on Man*, Epistle 4

What does the word "value" mean to you? In a business context, perhaps you define it as your company's share price, or the book value on your corporate balance sheet. If you're a marketer, you may think of it in terms of market share or customer satisfaction. As a parent, spouse, friend, or shopper, you may consider it a measure of something decidedly more personal.

With so many competing notions of value, and with the temptation to interpret value as whatever notion is most convenient for us at the time, how can anyone confidently talk about "managing for value"?

In the pages that follow, we offer a perspective on value and value creation that we hope clears away the confusion over these much-abused terms. What makes our definition of value different from the myriad of others is that our notion of value is not a social construct. It is not specific to time, place or context. It has nothing to do with anyone's opinion. It is an idea driven by nature through an instinctive, collective process.

Value creation has nothing to do with beliefs – yours, ours, or anybody else's. You may be familiar with the expression, *"Fifty million Frenchmen can't be wrong."* Well, yes they can. The same goes for 300 million Americans, a billion-plus Chinese, and so on. After all, it is entirely possible that every person working for Enron thought that their company was creating value when, in fact, it wasn't. Tens of thousands or even tens of millions of people believing that they are creating value does not make it so. Value creation, when properly understood, is not simply someone's ethical perspective on how to manage a company. Value creation is a self-generating, self-governing, basic planetary imperative based on nature itself, and if you don't uphold it, the planet will shut you down every time.

Consider a continuum of value where on one end we have the most basic of raw materials, and on the other, the consumers of these materials. Whether we're drilling for oil, pumping gas at the local service station, or driving the latest Jaguar, we are all participants somewhere within this value chain which rules the globe.

The reason it rules the globe is simple. Beyond our basic drives of food, shelter, and sex, we are driven to try to make each day of our lives a little better than the day before. To do so means finding ways to become happier, and that, at its most basic level, means taking the resources available to us and using them to create value. No matter what products or services we strive to create, our overall purpose is the same: delivering happiness to ourselves and creating ongoing value in our lives.

Value, in other words, is really just another word for happiness, at least from the perspective of the consumer. Consumption is a non-stop process that occupies every moment of our existence, whether we're conscious of it or not. Happiness is the cognitive experience that dominates our waking lives. As Richard Layard writes, *"We are programmed to seek happiness."*

However, he also writes that, *"Generally, what makes us happy is good for us, and has therefore helped to perpetuate the species."*[1] In other words, the creation of a system for delivering value – we call it business – was no accident. It was inevitable. We will say much more about this later, but for now, let's talk further about the basic value imperative by which we are all instinctively governed.

For humans, again, value equals happiness. To help deliver this happiness to ourselves, we at some point created businesses that could generate the products and services to make us a little happier each day. "Happier" might mean more comfortable. It might mean more excited or

---

[1] Richard Layard, *Happiness: Lessons from a New Science*, revised edition, (London: Penguin Books, 2011), p. 224.

interested. It might mean more entertained. It might mean better able to dust high window ledges. Business was the means of delivery. For any business to survive, not only must it deliver happiness that customers are willing to pay for (the *"cash applause of consumers"* as one observer puts it)[2], but the cash it receives must also be sufficient to cover that needed to pay its own suppliers, while at the same time ensuring a competitive return on the capital invested. From the perspective of business, we can therefore express value a different way. Later, we will introduce this other definition of value, and using it as a basis we will assert the concept of blue line management, the core thesis of this book. Blue line management is an approach that uncompromisingly focuses talent, energy, and decision-making on the sole objective of creating value. Every decision a company makes has an impact on value; it either creates value or destroys it. To put it in the starkest terms, blue line companies last because they are focused on long-term value. Other companies, which we refer to as red line companies, inevitably die because they are focused on other misguided definitions of value.

## We Want Our Stuff

Allow us to talk a bit more about the concept of happiness. Since humans created business for the purpose of delivering happiness, and since this book is about what we are calling the value creation imperative, which stems from humankind's overarching desire to be happy, happiness matters. A lot.

Happiness is relative, of course. What satisfies our needs, and therefore motivates us, may not satisfy yours. Some people may think of clean drinking water as their greatest need. Others might not feel happy unless presented with three different options for sparkling bottled water at a fancy restaurant. If you're a subsistence farmer in Southeast Asia and you survive the winter, you're happy. If you're a middle manager who hits the targets and gets that coveted promotion, you're happy. If you're Warren Buffett giving away 99% of your wealth (currently estimated at

---

[2]Deidre N. McCloskey, "A Kirznerian Economic History of the World", *The Annual Proceedings of the Wealth and Well-Being of Nations*, Upton Forum, 2010, p. 58.

$47 billion)[3] to good causes, and succeed in persuading a bunch of other billionaires to do likewise,[4] you're likely to be very happy. What you seek, what helps you survive, what motivates you, is happiness.

Since what makes us happy is different from what makes you happy, what enables happiness for all of us – and anyone else – is choice. We are not made happy by the same things, so we deeply appreciate the chance to pick and choose the sources of our happiness from among a great variety of possibilities. The massive growth of consumerism, particularly in the last century, has today given us unprecedented amounts of choice. There really are umpteen different types of bottled water to choose from. We can buy bacon-flavored chocolate, and possibly even chocolate-flavored bacon, and if you're in the market for a new cell phone, well, the options are seemingly endless.

In the 1980s, a Canadian doctor we know took a group of visitors from Leningrad (now Saint Petersburg) on a tour in Canada, including a visit to a regular grocery store. The Russians were astonished at the huge number of choices available, in particular the dozens of varieties of breakfast cereal. They commented that back home in Leningrad, they had but two. Their Canadian host pointed out that he really only liked two of the cereals on display, so he didn't understand the need for so many. But he realized that while other customers at the store might also enjoy only two of the cereals on offer, their two favorites would probably be different from his. Perhaps this crazy plethora of cereals was needed after all, at least according to the definition of our collective happiness.

A similar story was told by heavyweight boxing champion Vitali Klitschko on describing the shock he felt during his first visit to an American supermarket: *"I thought there was only one type of cheese, you know, the thing we'd always called 'cheese,' and in a grocery store, I saw a hundred kinds of cheese! It was amazing."*[5] Why a hundred cheeses? It's the only way to accommodate everyone's taste.

---

[3]T. Leonard, "The dinner that cost Bill Gates, Warren Buffet and other celebrities billions", *Daily Telegraph* online, 2010, http://www.telegraph.co.uk/news/world-news/northamerica/usa/7929657/The-dinner-that-cost-Bill-Gates-Warren-Buffett-and-other-celebrities-billions.html

[4]http://givingpledge.org/

[5]http://www.grantland.com/story/_/id/7128487/vitali-wladimir-klitschko

Naturally, it doesn't stop with cereal and cheese. We want choice for everything we desire. *"'Desire' as a fundamental aspect of the self,"* wrote historian Jan de Vries, *"is not a product of modern industrial capitalism; its origins are to be found earlier."* It began, he asserts, in Western Europe in the seventeenth century, during a process he termed the *"Industrious Revolution,"* when increased consumption of luxury goods led to a desire for more income, changing people's working habits and spurring the creation of faster, more efficient production methods.[6]

We could go back even further. What we desire has always driven our decisions regarding where and how to direct our time and energy. But as we'll see shortly, it is only within the last few centuries that an efficient, shared method for really improving our lives has emerged, changing the landscape forever.

So what does all this conversation about happiness and choice have to do with you as a business leader? Everything. As consumers, we value, and are therefore willing to pay for, the products and services that make us happy, whatever our definition of happiness is. The only job of any business is to figure out what makes people happy and then try to deliver that happiness at a price the consumers find reasonable, while at the same time earning a competitive return on invested capital. If a business can do that sustainably, it is value-creating.

But if it can't find out what makes people happy, or if it thinks it has figured it out but is wrong, or if it figures it out but charges too much for that happiness, or, finally, if it has figured out what makes people happy and charges the right amount for it but can't make enough profit to adequately compensate for its own capital outlay, it will die. This may take some time, but the inevitable will happen eventually. We as consumers will kill the business because it is not creating value in our lives.

Sometimes we don't know what we want of course, and that can make your job very tricky. When Gillette released their MACH3 three-blade razor in 1998, it was an enormous success. Soon after, the company surveyed its customers to see if they might prefer a four-blade razor. The majority said no, they were quite content with the three blades, so Gillette decided not to develop a new razor.

---

[6]J. De Vries, *The Industrious Revolution: Consumer Behavior and the Household Economy, 1650 to the Present*, (New York: Cambridge University Press, 2008), p. 43.

In 2004, rival shaving company Wilkinson Sword (Schick in the US) released a four-blade razor, the Quattro. Consumers embraced the Quattro, buying it in droves. Had Gillette's customers lied? No – they simply didn't know what they wanted. (This particular battle is far from over. Gillette introduced the five-bladed Fusion razor in 2006.)

Not only are there differences between what makes us happy and what makes you happy; even more important, what makes any of us happy today may not make us happy tomorrow. As shown above, we don't even know whether we prefer three razor blades or four. How can we be expected to decide on the really important stuff? Did we know we wanted the iPad before it arrived? Did we know we needed cameras in our cell phones? No.

The only thing certain is that companies need to constantly innovate in order to keep providing us with the things that make us happy, and therefore create value. As Steve Jobs explained, *"It's not about fooling people, and it's not about convincing people that they want something they don't. We figure out what we want. And I think we're pretty good at having the right discipline to think through whether a lot of other people are going to want it, too. That's what we get paid to do."*[7]

It's hard enough for companies to figure out what makes us happy when we ourselves don't always know what we want. Making matters worse, market researchers and others who research the customer psyche can't necessarily be relied upon to accurately reveal what we find valuable. The French vilify McDonald's, yet there are now over one thousand McDonald's, or McDo, as they are called locally, in France – including one in the Louvre. Even worse for the naysayers, the McDonald's on the Champs-Élysées is the most profitable in the world.[8]

Behavioral researchers have begun to question whether all of this choice is really good for us – whether, in the grand scheme, it really serves to create value for humanity.[9] According to one observer, *"Choice*

---

[7]http://www.investinganswers.com/a/50-quotes-genius-behind-apple

[8]http://www.huffingtonpost.com/2009/10/05/mcdonalds-to-open-a-resta_n_309453.html

[9]"The Tyranny of Choice", *The Economist*, December 18, 2010, pp. 111–113.

*no longer liberates, but debilitates. It might even be said to tyrannise."*[10] It is suggested in some circles that such a bewildering array of choice in just about everything we buy leads to confusion, indecision, panic, buyer's regret, and anxiety. Researchers at McKinsey & Company estimate that if one were to add up all the different sizes, shapes, colors, and flavors of all the products on offer in a major economy such as New York or London, it would come to over ten billion distinct items.[11]

Does this panoply of choice cause us to feel overwhelmed at times? Sure. But it's the unavoidable by-product of human innovation driven by human desire for happiness, that is, value.

Consider for a moment the alternative. Humans have been struggling for millennia to reach a state in which we ourselves, and not others, get to make the critical choices in our own lives. What's more, since our wants and desires are so varied, the only way they can all be accommodated is by the vast range of choice we are only now beginning to witness – no doubt there is still a long way to go. For practically the whole of human existence, most people on the planet had no choice but to defer to their social betters in terms of what level or status they could aspire to – or even, more simply, what stuff they were entitled to get. It's a good bet that our ancestors, observing our lives today, would have little sympathy for the plight of having too much choice.

Trader Joe's is a fast-growing retail grocery chain in the US The average Trader Joe's carries only about 10% of the stock-keeping units of a typical supermarket, and most of those products carry one of the company's own brands. For example, if you're looking for Frito Lay corn chips, you won't find them at Trader Joe's. You will find instead a store brand of pita chips made in a Frito Lay factory. If you're seeking a full range of snack foods, or branded products in any food group, you'll have to visit a more conventional grocery store.

Yet Trader Joe's is one of the fastest-growing retailers in the US. Friendly, high-quality service is one reason. But another is that some shoppers apparently prefer the limited range, preferring to ask for guidance

---

[10] Ibid., p. 112.

[11] Tim Harford, *Adapt: Why Success Always Starts with Failure*, (New York: Farrar, Straus and Giroux, 2011), p. 3.

from the merchandise buyers at Trader Joe's instead of making decisions themselves.

But there's an important point here that we shouldn't forget. Shopping at Trader Joe's is itself a choice. Outsourcing some of your shopping decisions to the people at Trader Joe's? That's a choice too, and a pretty powerful one. Should the habitual Trader Joe's clientele occasionally duck out to a Safeway or Winn-Dixie to get that must-have brand that they can't find at their favorite grocer, well, that's a third choice. Yes, we can rant all we want about the tyranny of choice, but again, don't try to play that card with your subsistence-driven ancestors who would probably want to stick a mastodon bone in your eye.

For business managers, the practical consequence of all this choice is simple, yet challenging. Choice is, as we said, about giving customers what they want, for a price they are willing to pay, while making a profit.

Let's focus on the first two parts of that statement: giving customers what *they* want, for a price *they* are willing to pay. You don't dictate these choices; they do. Every decision, from initial product development to final packaging, must be made with a view toward ensuring that the cost of adding features is less than what the customer is willing to pay. If you can accomplish this, you are reaching blue line nirvana: increasing both the customer's happiness and your company's value.

You can only deliver benefit to customers in two ways: by providing higher quality products and services – that is, increased happiness – at the same cost, or by providing the same products and services at a lower cost – increased happiness by a different name. Both equate to a bump-up in happiness without a corresponding increase in the resources required to deliver it. Value creation, and therefore the long-run survival of your business, depends on the achievement of at least one of the two versions of this feat.

Business history is littered with stories of companies that suffered because they failed to heed this lesson. Xerox insisted on adding more and more features to their copiers without any regard to whether these features were the ones their customers saw as important. When Canon then entered the fray with simpler machines that did what the customers really wanted them to do – namely copy documents well – without any bells and whistles, Xerox sales collapsed. As Peter Drucker reminds us, *"Quality in a product or service is not what the supplier puts in. It is what the customer gets out and is willing to pay for. A product is not quality because it is hard to make and costs a lot of money, as manufacturers typically believe. This is incompetence. Customers pay only for what is of use to them and gives them value. Nothing else constitutes quality."*

We are particularly fond of Toyota's approach to value. The Toyota Production System, as described by Professor Jeffrey Liker, *"starts with the customer by asking, 'What value are we adding from the customer's perspective?' Because the only thing that adds value in any type of process – manufacturing, marketing, development – is the physical transformation of that product, service, or activity into something the customer wants."*[12] This philosophy points to the almost religious significance Toyota places on squeezing waste from the system anywhere it can. Toyota believes steadfastly that every activity the company performs must contribute to value for the customer. Everything else counts as squandered resources.

As we said above, what makes us happy can also be what helps us survive. Our "stuff" is more than just expensive water in restaurants or razors with extra blades, to be sure. It's also what keeps us warm, feeds us, or helps make us better when we get sick.

This wasn't always the case, of course. Human history suggests it's only relatively recently that we started to make a noteworthy dent in the problems of basic survival, let alone becoming the technologically-indulged creatures we are today. For reasons we'll explain in a bit, true value creation is a relatively recent human phenomenon, with virtually all of the improvements in our living standards taking place over the last 400 years.

Sustained value creation is even more recent than that. The Industrial Revolution may have begun in earnest sometime around the middle or latter part of the eighteenth century, but it didn't translate to overnight value creation. Noticeably improved living standards among the working

---

[12]Jeffrey K. Liker, *The Toyota Way*, (New York: McGraw-Hill, 2004), p. 9–10.

classes were not evident for at least another 75 years. We haven't been getting our stuff for very long.

## Centuries of Subsistence

*"Throughout recorded history, most people in Europe – as elsewhere in the world – had possessed just four kinds of things: those they inherited from their parents; those they made themselves; those they bartered or exchanged with others; and those few items they had been obliged to purchase for cash, almost always made by someone they knew."[13]*

– Tony Judt, historian, 1948–2010

Until a few hundred years ago, the human experience had changed very little. Over the millennia since we first became upright, people reliably died young, cold, hungry, and of what we now think of as trivial diseases. This was a function of how human groups evolved, from hunter-gatherer bands of 10 to 100 people to larger tribes, chiefdoms, and eventually empires. As the size of our groups increased, we went from being able to share our catch and fires with our nearest kin, in what one scholar calls "evolutionary egalitarianism,"[14] to being part of a hierarchical system that concentrated economic power among a relative few. This left the rest of the population impoverished, and with negligible chance of changing their lives for the better.

Until we could find a way to circumvent this problem, we were stuck with subsistence, a condition that afflicted much of humanity well into the nineteenth century. In France and other parts of Europe, people even developed a form of hibernation, where they would virtually shut down their existence for half the year, focusing strictly on staying alive. They had nothing to do, no fields to till, and not enough food to sustain them if they were to go out and be active during the cold months. Writing in

---

[13] Quote taken from Tony Judt, *Postwar: A History of Europe since 1945*, (New York: Penguin Books, 2005), p. 337.
[14] M. Shermer, *The Mind of the Market*, (New York: Henry Holt and Company, 2008), p. 18.

the *New York Times*, Graham Robb pointed out that *"Villages and even small towns were silent, with barely a column of smoke to reveal a human presence."*[15] He also cited a 1900 edition of *The British Medical Journal* describing how peasants in Pskov, Russia, similarly slept for half the year: *"At the first fall of snow, the whole family gathers round the stove, lies down, ceases to wrestle with the problems of human existence, and quietly goes to sleep. . . . After six months . . . the family wakes up . . . and goes out to see if the grass is growing."*[16]

## A Matter of Power

The past was a simpler time undeniably, but undeserving of the nostalgia often ascribed to it. Many people in the old days had more "leisure time" than we have today, but when we speak of leisure in this context, it's far from the idea of leisure we have today. It's not the time we take to pursue cultural, sporting, and social activities. Well into the nineteenth century, the great majority of people on earth lived lives that were little better than those of our Stone Age ancestors. Often lacking the calories needed for a full and productive life and the consumer goods that offer pleasure and comfort, they did little more than survive.

Sometimes, not even that. Even in relatively rich countries like France, large swathes of the population were never more than one bad harvest away from famine. Indeed, the lingering effects of this threat and the persistent fear of hunger could be found in the many proverbs and ritual phrases in use well into the twentieth century: *"Don't eat everything at once," "You've got to stretch things out,"*[17] and so on.

Even as conditions improved, entrenched habits of conservation and frugality died hard. Historian Eugen Weber noted that many peasants in France around the turn of the twentieth century continued to eat inferior barley bread even as white bread became more plentiful and less expensive.

---

[15] G. Robb, "The Big Sleep", *New York Times*, 2007, http://www.nytimes.com/2007/11/25/opinion/25robb.html

[16] Ibid., http://www.nytimes.com/2007/11/25/opinion/25robb.html

[17] Eugen Weber, *From Peasants to Frenchmen: The Modernization of Rural France, 1870–1914*, (Stanford (CA): Stanford University Press, 1976), p. 137.

They feared they would eat the white bread *"with too much pleasure, and hence consume too much."*[18]

In the Netherlands, old habits linger even to this day, despite the fact that the country has for centuries been one of the world's richest. The Dutch are still reluctant to light candles until after sunset, evoking a cultural memory in which saving candles mattered. What's more, they still re-use teabags and coffee grounds as a matter of course.[19] Tony Judt reminds us that for *"the overwhelming majority of the [west] European population up to the middle of the twentieth century, 'disposable income' was a contradiction in terms."*[20] It's jarringly easy to forget just how recently the grinding routines of material scarcity held sway over every aspect of human life.

People were less economically active in pre-modern societies in part because there was so little to buy – assuming one could even get one's hands on some money. Why was there so little stuff available? Two reasons. First, in most societies, multiple unseen forces were marshaled against change to the prevailing social order. Second, and critically for this book, there was no mechanism to enable the people with the ideas and inventions to help us survive and make us happy – today we call them entrepreneurs – to get the financial backing they needed to see their ideas to fruition.

Put more simply, there was no way to make sure we got our stuff. Historians tend to ignore the role consumers played in human development, but we see it differently. Consumerism, by spurring the desire for us to earn money, was powerful enough to subvert a traditional hierarchy that had lasted for centuries and generate an endless upward spiral of improvement in the human condition as entrepreneurs and companies became able to provide us with what we needed to make our lives better.

What sort of forces had conspired to keep humankind in servitude? For many centuries, entrepreneurs, at least the sort that tried to com-

---

[18] Ibid.

[19] Deirdre N. McCloskey, *The Bourgeois Virtues: Ethics for an Age of Commerce*, (Chicago: University of Chicago Press, 2006), p. 425–426.

[20] Tony Judt, op cit., p. 337.

mercialize innovations and make them available to the wider public, were frowned upon. There are only two ways to make money. The first is to improve the world and take a cut in the process. The second is to rip people off. Let us call the first method Productive Entrepreneurship and the second Conscious Fleecing. For a long time, it stood to reason that Productive Entrepreneurship was not viewed as a reasonable path to riches or status. Economist William Baumol pointed out that for the Romans, *"As long as it did not involve participation in industry or commerce, there was nothing degrading about the wealth acquisition process."*[21] Those who did acquire their wealth via industry or commerce were typically freedmen – former slaves – and therefore socially stigmatized.

In medieval Europe, it wasn't that enterprise was frowned on; it was merely considered a waste of time unless it helped promote warfare or aided in capturing a neighbor's castles and lands. Ideas for better siege machines or more sophisticated weaponry had a good chance of seeing the light of day, but those ideas aimed at improving the lot of the common man made little headway.

The ancient Chinese had a similarly unenthusiastic view of commerce. Instead of inventing things and working to make everyone's lives better, thousands of men sought advancement by sitting the imperial examinations and becoming bureaucrats. If they passed, they moved into a position of power with access to tax and other legal and not-so-legal revenues. Even in the twenty-first century, such behavior is not uncommon. Practices similar to those of the Chinese mandarins have emerged in Russia, where government officials – dubbed *"bureaucrat-entrepreneurs"* by *The Economist* – exploit a weak Russian state through a combination of racketeering and outright theft of budget revenues.[22]

Throughout history, sumptuary laws – laws regulating consumption – were also used to restrict what certain people could buy or wear, therefore maintaining social rank, privilege, and discrimination. The Romans had

---

[21] W.J. Baumol, "Entrepreneurship: Productive, Unproductive, and Destructive", *Journal of Political Economy*, 1990, 98(5), part 1., p. 899.

[22] "Frost at the Core", *The Economist*, December 11, 2010, pp. 25–28.

rules about how many stripes you could have on your toga and who was allowed to don silk. In ancient Greece, only prostitutes could model embroidered robes in public. In Imperial China, only those of noble blood were permitted to wear yellow. Even in late nineteenth-century France, the bourgeoisie and intellectual classes frequently expressed contempt for those peasants and working-class folk who tried to emulate the dress of their social betters.

But history teaches us something else: that laws or social norms designed to try to control what people can do or get, don't work. Human ambition and the drive for entrepreneurship will come to the fore time and time again. Did Prohibition in the United States work? Of course not. Creating rules that prevent people from getting access to the stuff they want doesn't stop them from wanting it. Or from finding ways to get it.

The problem throughout all these generations was that though entrepreneurship is a relentless human drive, it requires a system to make it work. While it is romantic to think that money doesn't make the world go round, it does. If innovative ideas can't get funded, they can't transform from ideas into reality – they never become the stuff that makes our lives better.

For thousands of years, the chances of an entrepreneur realizing his dreams were extremely limited. Capital allocation – funding – was driven purely by relationships. Because assessing the true value of an idea was so difficult, investors made decisions based not on whether an idea had merit but whether they thought the person behind the idea was trustworthy.

But there was a larger problem. For most of our history, in most places where we existed, the wealth was held by a select few, and these few were very hard to reach, unless you had the right contacts. For those with innovative ideas, procuring the funding necessary to create their new machine or product was next to impossible. The investment environment was, in other words, extraordinarily inefficient. The funds that were available systematically went to the wrong people because they were the relatives, friends, or friends of friends of those in power; or because they were politically shrewd; or because they were good at passing exams; or because they were good at killing more people than the next man, or perhaps the right people at the right time.

In such environments, the vast majority of new, potentially worthwhile ideas failed to acquire the resources they needed. The result? Thousands, possibly millions, of opportunities to improve the lot of humankind remained in the shadows. In the pre-modern world, personal connections and the circumstances of one's birth mattered much more than the inherent value of an idea.

Not only did the possibility of getting money for your idea depend mostly on who you knew or were related to, even if you somehow got your hands on some funds, the risks of borrowing were big. From the moment the Babylonian King Hammurabi announced his famous codes of law nearly 4,000 years ago, being in debt became a seriously dangerous proposition. If the debtor wasn't able to pay off what he owed, the lender was entitled to three years hard labor from him – although in fairness to Hammurabi, the debtor did have the option of sending his wife or child instead.

It wasn't just a matter of paying back what you had initially borrowed. Interest rates on personal loans were huge because the risks of lending were so high. First, there was the possibility that the idea might not work. Second, books like the Bible, the Torah, and the Qur'an all spoke out against lending and charging interest, calling it usury (which still exists today under Islamic law), meaning that those who lent money could not legally charge interest.

There was a way around this however. Expressions such as *"Thou shalt not lend upon usury to thy brother"*[23] could be interpreted to mean that interest *could* be charged on money lent to other tribes, for instance by Jews to Gentiles – hence the *banchi* or benches on the Rialto in Venice, where Jewish moneylenders operated and where Shakespeare's Antonio (*The Merchant of Venice*) came to borrow money from Shylock. Jews who charged interest on loans to fellow Jews faced social exclusion. That was bad enough, but the moneylenders also risked their lives lending to non-Jews, because as a minority, they were that much more vulnerable to persecution, or even death, should things not go smoothly. There was no institution of law to protect lenders, contributing again to

---

[23] King James Bible, Deuteronomy 23:19.

what was a wildly inefficient, illogical system. Small wonder interest rates were astronomical.

If borrowing privately was fraught with risk, trying to borrow from a bank wasn't much better. The banks had long operated on a simple, ancient business model: those who managed to scrape together enough savings could give the bank some money to protect it, as long as they were willing to pay for the favor. The banks' attitude was this: just as a primitive man might try to save some corn over the winter but would naturally expect there to be less of it when he went back to it later because of climate, other hungry animals, and so on; surely depositors would not expect all of their money back after depositing some of it for protection?

Quite secure in this logic, the banks would lend the deposited money out to those it felt were better able to repay it – typically, their friends, or those who didn't really need the money in the first place. The banks were far less than the efficient facilitator of funds society needed; they were part of the problem instead.

A further impediment to the realization of entrepreneurial ambition was the absence of "free incorporation." Until the nineteenth century, incorporation in most countries required the consent of the ruling elites – the monarch, parliament, or whoever else sat at the head of the pro-verbial table. In England, for example, not until 1844 could corporations be formed without state permission. Not surprisingly, incumbent corporations used these rules to their advantage by limiting competition and maintaining a bigger share of the pie.[24]

Finally, for those entrepreneurs who actually gained an opportunity, the decision to pursue it wasn't so easy. We all know the experience of hoping for the chance to do something only to feel extremely intimidated and anxious once that chance presents itself. Entrepreneurship in any age carries risk, and for many in the pre-modern world, the risk was too great. As the American historian Eugen Weber reminds us in his work on late nineteenth-century French peasantry, *"The narrower the margin, the*

---

[24]Timur Kuran, *The Long Divergence: How Islamic Law Held Back the Middle East*, (Princeton: Princeton University Press, 2011), p. 121 and p. 319 (n.13).

*less the chance of experiment. Only the rich took chances – or the irresponsible."*[25]

In short, if the poor wanted to survive, they had little choice but to work with what they had and think no bigger. Even if you did have a great idea for changing the world, was it worth investing the time and energy, and assuming the risk, to try to make it work? Probably not. Even for the boldest or most innovative thinkers among the poor (i.e., most people), to try and fail didn't just mean dusting yourself off and trying again; in most cases it meant starving, and quickly.

Weber goes on to say that *"Subsistence farming – raising a bit of everything and making one's own bread and clothing – was a matter not of blind routine but of calculated necessity."*[26] This autarkic existence made it virtually impossible for peasants to accumulate any surplus or savings. In those instances where a bit of accumulation was possible, it was swiftly taxed away, seized by rapacious lenders charging usurious interest, or spent on adjoining land. Precious little was left for trying to make the world a better place.

## The Rise of the Consumer

*"Every man is a consumer, and ought to be a producer. He fails to make his place good in the world, unless he not only pays his debt, but also adds something to the common wealth. Nor can he do justice to his genius, without making some larger demand on the world, than his subsistence."*
> – *The Conduct of Life*, Ralph Waldo Emerson,
> American philosopher and poet, 1803–1882[27]

As Emerson so sensibly points out, our roles as consumers go hand in hand with our roles as producers. We produce things to consume; and because we want to consume, we produce.

---

[25] Eugen Weber, op. cit., p. 480.

[26] Ibid., p. 481.

[27] http://classiclit.about.com/library/bl-etexts/rwemerson/bl-rwemer-conduct-3.htm

This wasn't always the case. Renaissance Europe may have had many fine things – art, architecture, gold, jewels – but it wasn't what we would today consider a consumerist society, at least not in the collective sense. Luxuries were the exclusive domain of the aristocracy – the landed rich – and remained to the common man as out of reach as flying to the moon. Because of this vast gulf between the haves and the have-nots, our subsisting, hibernating peasants felt no incentive to work harder in the fields than they already did, or to make wares in their spare time to sell, since there wasn't much they could spend any extra money on, anyway.

Still, they felt the desire to reach beyond, to do more. They felt the powerful urge to produce and invent – to somehow enhance the comforts of their life and find ways to bring what is sometimes called "old luxury" within the grasp of more than just a privileged few. Though they may not have described the urge this way, they were striving to create "new luxury" that could be available, ideally, to everyone.

We can trace the beginnings of what we think of today as consumerism – a system that encourages the purchasing of goods and services in ever-greater amounts – back to the seventeenth century. The clearest evidence of these beginnings can be seen in the trading activities initiated by the Dutch and British East India companies. When their ships started returning from their overseas voyages, they brought back with them products and materials – spices, coffee, tea, silk, porcelain – that were available to anyone who had the means. As a result, people were encouraged to find new ways to make money. Even women and children were allowed to participate in the new system, which has been termed proto-industry.

Suddenly, from having no incentive to earn more money or produce anything different – or having much spare time to do it anyway – people had a reason to use any time not spent surviving, to try to make more money so they could buy the fantastic things coming back from unknown, faraway places. The result was predictable and remarkable. The output of the average person rose dramatically and only in part because of technological progress. This change came about largely for a much simpler reason: people wanted to earn more so they could buy things. It was a simple but powerful chain of logic. Because these things were being produced in large quantities for the first time, they were becoming available to anyone who could make enough extra money to buy them. This opportunity – the opportunity to consume – encouraged people to work their tails off.

Was their desire to buy blind, aimless, or irrational? Hardly. People wanted to buy the things that the ships were bringing in because they were acutely aware that those things would improve their lives. In England, for instance, the arrival of lightweight calico from India and gingham from the Far East meant people could discard the traditional, home-grown thick linens and wools in summer. In Holland and else-where, the arrival and quick dissemination of tea transformed it from being a drink of the elite to a universal comfort and pleasure. People suddenly had ways to access the lifestyle of the previously untouchable rich. They drank coffee, smoked tobacco, ate chocolate. It made them feel good. It made them feel important.

And, once they got the taste for it, it made them want more. The Dutch East India Company at first imported a few thousand plates, bowls, vases, and the like from Asia. By the end of the eighteenth century that number had grown into the millions, and because of this mass produc-tion and mass availability, fewer and fewer items remained as the exclu-sive domain of the elite class. As consumerism exploded, social distinction eroded. What might be sold to the aristocracy as a high-quality, high-priced product was soon being produced in volume and sold cheaply to the masses. Jan de Vries traced this pattern by assessing the different quality of Delft tiles found in canal houses and more humble abodes throughout Holland. While the rich continued to enjoy the most expen-sive wares produced in Delft's workshops, various gradations of quality (and the resulting lower prices) made it possible for others to enjoy the decorative and aesthetic pleasures that came from owning such products.

The pattern continues today of course. Cheap versions of the latest Chanel bag or Hermès scarf are never far away. Even the iPad didn't last long as the only tablet on the market, with Samsung, Amazon, LG, and other manufacturers rushing to offer lower-priced versions and capture those consumers unwilling, or unable, to pay Apple prices.

Commentators from Jean-Jacques Rousseau in the eighteenth century to John Kenneth Galbraith in the twentieth lamented these developments, equating mass consumerism with an exploitative form of capitalism that loses sight of the greater good and creates an underclass that can't distinguish between what it wants and what it needs. This, they claim, leads people to want to emulate those socially higher than themselves, rather than support innovation that truly meets their needs and desires. Thus, the argument goes, this type of consumerism must be bad for people.

But as the great economist Joseph Schumpeter wrote, *"The capitalist achievement does not typically consist in providing more silk stockings for queens, but in bringing them within the reach of factory girls in return for steadily decreasing amounts of effort."*[28] Three centuries of innovation spurred by the need to provide what consumers want has given us not only cheaper commodities and all the gadgets that help us organize our time and maintain our social networks, but also the telephone, MRI scanners, and key-hole surgery. It is of course true that some of the things we come up with don't provide happiness or create value, but these inventions don't last. Eventually, enough of us reject them and they go away.

Alleviating the burden of survival did more than allow people to breathe more easily. It also gave them the chance to think – to focus outside their immediate environment, to consider society as a whole, the way it worked, what it might do better. Without a drastic change in the way society functioned, from feudalism to greater self-determination, and with what economic historian Joel Mokyr referred to as *"a conscious belief in the possibility of continuous betterment of society,"*[29] we would never have had the Age of Enlightenment of the eighteenth century. In addition, humankind might never have questioned the way societies operated and we might never have developed the scientific method. Consumerism also gave the English, Napoleon's so-called "nation of shopkeepers," an economic strength that enabled campaigners to establish the political will to abolish slavery. Our desire for happiness is a powerful collective force. It weeds out the ideas that don't truly create value and supports

---

[28]Thomas K. McCraw, *Prophet of Innovation: Joseph Schumpeter and Creative Destruction*, (Boston: Harvard University Press, 2007), p. 9.

[29]Joel Mokyr, *The Enlightened Economy: An Economic History of Britain 1700–1850*, (New Haven and London: Yale University Press, 2009), p. 33.

the ones that do. The drive to create stuff that makes our lives better comes not from a selfish, aimless, or counterproductive instinct but from a communal sense of value that over time, impels economic, social, and political change in ways that few people, if any, understand.

In order to have the things we want, we need people and businesses to have good ideas, be able to convert those ideas into stuff, and then to figure out a way to get that stuff to us. This has always been the necessary equation for the creation of value, but until a few hundred years ago, it wasn't possible. Whether our most pressing wants are represented by basic survival, Asian porcelain, or gold-plated faucets, each part of this chain – the original idea, the ability to turn it into something, and a method of delivery – must be satisfied. But consumerism couldn't become a reality until a mechanism surfaced that would enable it. It's time to talk about what kick-started society into the explosion of innovation that makes our lives today longer, easier, and happier than those of our ancestors. It's time to talk about the capital markets.

# CHAPTER TWO

# THE GLOBAL CAPITAL MARKET

*"In 1750, most people would probably have said that the pre-industrial configuration of the world's economy was largely a permanent state of affairs. That the world had always been like that and probably always would be and they would have had the facts on their side."*[1]

– Michael Spence, Nobel Economics Laureate, 2001

We've argued that, in order for the gears of consumerism to kick into motion, three things were necessary: first, good ideas; second, time and resources for the entrepreneurs with the ideas to develop them into useable stuff; and third, a way to get that stuff to market and ultimately, into the hands of consumers. We've argued further that it is our role as consumers that has driven the massive social and political changes of the last 400 years. If the difficulty of borrowing money to start a business or develop an idea was a major force in keeping humankind at a level of subsistence, then some seismic shift must have occurred that led to today's circumstances where we clearly get our stuff by the bucketful.

So what changed? What strange mechanism emerged to ensure capital could be allocated efficiently and fairly? What system arose that could somehow naturally and objectively direct the right funds to the right people? At one point in our existence, getting funding for good ideas was nearly impossible. The risks were enormous and the rates excessive.

---

[1]Michael Spence, *The Next Convergence: The Future of Economic Growth in a Multispeed World*, (New York: Farrar, Straus and Giroux, 2011), p. 15.

Something happened to get the Cost of Funding (COF) down to a level at which investors could lend money without risking everything they owned, and entrepreneurs could accept the loans without signing away their lives as collateral.

The initial action that started moving money into the hands of those who could do something useful with it, was a small experiment launched by the Dutch in the fall of 1606. The direct result of this action is the rise in standard of living and extended life expectancy we all enjoy today. But we'll get to that, since it didn't just happen overnight. Before this monumental brainstorm, there were a number of false starts.

One example of this was to drive innovation through attractive incentives. For instance, we've long depended on ships to transport raw materials and tradable goods. But for millennia, shipping was a treacherous method of transportation for a variety of reasons, not least trying to navigate vast bodies of water to get from one point on the map to another. Errors were common, as was the resulting frequency of wrecks, lost ships, and dead crew.

The main issue was a notoriously resistant beast called longitude. For centuries, measuring how far a ship had traveled east or west from a fixed point proved sailing's most difficult test. In 1675, King Charles II of England set up the Royal Observatory in Greenwich to try to solve the thorny longitude issue. The idea behind the Observatory was to produce charts to help sailors find their longitude by tracking the position of the moon relative to various stars. Since timepieces in the seventeenth century were still inaccurate and unreliable, this idea seemed the best alternative, but it proved of little use. In 1714, the British Government increased the incentive to willing entrepreneurs by announcing what became known as the Longitude Prize, whose reward was the immense sum of £20,000 to anyone who could come up with a method to assess longitude to within five-tenths of a degree. Clockmaker John Harrison eventually invented a timepiece that was so accurate that the judging committee declared his result a fluke and refused to award the money – at least not until the King intervened, at which point they found it prudent to cooperate.

In a similar vein, Napoleon offered a prize for innovation in food preservation for his army, leading to the development of modern canning. And the Orteig Prize spurred Charles Lindbergh to make his transatlantic flight.[2] Although not a prize as such, it's fair to say that President John F. Kennedy's exhortation in 1961 that *"This nation should commit itself to the goal, before this decade is out, of landing a man on the moon and returning him safely to earth,"*[3] had a similarly galvanizing effect to that of the Longitude Prize, and led more or less directly to Neil Armstrong setting foot on the moon in the summer of 1969.

More recently, the X Prize Foundation, whose mission is *"to create radical breakthroughs for the benefit of humanity, thereby inspiring the formation of new industries, jobs and the revitalization of markets that are currently stuck,"*[4] has offered a series of multi-million dollar prizes for specific innovations in space travel, automotive propulsion, and human genomics.

The problem with these types of incentives and prizes is twofold. First, they are time-specific and self-contained. Second, they serve to orient vast amounts of effort and energy around objectives defined by the few; they are not part of a larger, ongoing mechanism that aids a continuous flow of funding to those with valuable ideas for moving humanity forward. Prizes can thus produce particular results at specific times, just as sporadic impulsion from government leaders can spur accomplishments – who doesn't still get misty-eyed at the moon landing? – the inventions and advances these practices generate are unpredictable and unreliable. In the grand scheme of human endeavor, they are drops in the ocean. It is another event that initiated and sustained the constant stream of innovation we witness today. By the time the British Government

---

[2]The Napoleon and Lindbergh examples are from "Prizewinning Policy," by Annie Lowrey, *Slate*, December 27, 2010. See http://www.slate.com/toobar.aspx?action =print&id=2279272. The Orteig Prize was a $25,000 reward offered in 1919 by New York businessman Raymond Orteig to the first aviator(s) to fly non-stop from New York City to Paris or vice versa. On offer for five years, it attracted no competitors. Orteig renewed the offer in 1924, which Lindbergh claimed three years later.

[3]http://www.jfklibrary.org/Historical+Resources/Archives/Reference+Desk/ Speeches/JFK/Urgent+National+Needs+Page+4.htm

[4]http://www.xprize.org/

announced the Longitude Prize, this event had already come to pass, and it would have a much more dramatic, far-reaching, and long-lasting effect: a group of Dutchmen decided to share ownership of their company with anyone willing to pay for it.

In 1602 shares in the Dutch Vereenigde Oost-Indische Compagnie (VOC, better known as the Dutch East India Company) were issued, suddenly creating what is usually considered the world's first publicly traded company. While commodity exchanges had existed in various forms since early civilization (the UK is dotted with ancient Corn Exchange buildings that are now used as art and entertainment centers), and brokers trading in bank debts had plied their trade since at least the twelfth century, it wasn't until this event that company ownership truly changed and the common man saw opportunities he had never seen before.

There are other claimants to the title of first public company, including a twelfth-century water mill in France and a thirteenth-century company intended to control the English wool trade, Staple of London. Its shares, however, and the manner in which those shares were traded, did not truly allow public ownership by anyone who happened to be able to afford a share. The arrival of VOC shares was therefore momentous, because as Fernand Braudel pointed out, it opened up the ownership of companies and the ideas they generated, beyond the ranks of the aristocracy and the very rich, so that everyone could finally participate in *"the speculative freedom of transactions."*[5] By expanding ownership of its company pie for a certain price and a tentative return, the Dutch had done something historic: they had created a capital market.

Not surprisingly, the idea of shared company ownership quickly caught on and spread. What began in Amsterdam soon moved to the Dutch colony of New Netherland in Colonial America, which included a little island called Manhattan. Share-trading also began to catch on in England and elsewhere in Europe, and started to travel further around the world, reaching Bombay, Hong Kong, and Tokyo by the second half of the nineteenth century. During those 250 intervening years of propagation, capital markets had generated enough money to help set in motion another significant event, the Industrial Revolution.

---

[5] F. Braudel, *The Wheels of Commerce*, (New York: Harper & Row, 1982), p. 101.

Thanks to the Dutch, capital markets had taken root and quickly expanded. But the value creation imperative – the natural process by which the right ideas are enabled and the wrong ones rejected – would not perfect the process for some time, because certain inherent issues needed to be purged first from the overall system.

One serious glitch in the early days was that shareholders were liable for *all* of the debts of the companies in which they held shares, proportional to the number of shares they owned. The great grandfather of Scottish novelist John Buchan was a shareholder of the City of Glasgow Bank in the mid-nineteenth century. Bad loans and speculative investments, coupled with a certain amount of fraud, caused the bank to fail spectacularly in 1878, with debts of over £6 million. The company's shareholders were liable for its debts, which meant many of them had to pay far more than the value of their original shareholding – including Buchan's ancestor who lost everything.[6]

If the older Buchan had won his appeal to have the Companies Act applied to his case, things wouldn't have been so bad. This act extended limited liability to all companies in England in 1856 (the East India Company had had limited liability status since 1662), which effectively assured shareholders that, even if the company in which they had invested were to go belly-up, they would lose no more than the value of what they had originally paid. Thus, one snag in the original design of the capital markets was fixed, and a boost to growth was the outcome. As William Bernstein put it, *". . . limited liability is a near-absolute requirement for healthy public participation in company ownership; without it, the public will not supply equity capital to growing companies."*[7]

Another major hindrance was debtor's prisons. Made infamous in the work of Charles Dickens, debtor's prisons hardly made it easier for people to pay back their debts, not least because it's hard to earn a paycheck while you're in jail. The cramped conditions in such places also

---

[6] James Buchan, *Frozen Desire: The Meaning of Money*, (New York: Farrar, Straus and Giroux, 1997), pp. 208–210.

[7] W.J. Bernstein, *The Birth of Plenty: How the Prosperity of the Modern World was Created*, (New York: McGraw-Hill, 2004), p. 150.

generated the constant threat of death from transmissible diseases. Impris-
onment for debt was abolished in the United States in 1833, and most
European countries followed suit shortly thereafter. Another obstruction
to high-functioning capital markets was removed, leading to another
upturn in growth.

As the normal inborn snags were detected and dealt with, the capital
market boomed, emerging from its early growing pains to become truly
global. Today, there are nearly 250 stock exchanges all over the world,
from the mighty New York Stock Exchange, with some 2,800 listings and
a market capitalization of nearly $12 trillion,[8] to the 19 stocks[9] traded on
the Port Moresby Stock Exchange in Papua New Guinea. One of the
newest to join the bandwagon is the Rwanda Stock Exchange, which
opened in Kigali in January 2008, trading exclusively in government
bonds at first and expanding, in 2009, to equity listings. The World Fed-
eration of Exchanges reported that by year-end 2012, total market capitalization for the 52 regulated exchanges it represents was approximately
$55 trillion, with more than 46,000 listed companies.[10] We seem to know
when to take a good idea and run with it.

## How Capital Markets Help Us Get Our Stuff

In pre-modern times, not unlike today, there were entrepreneurs – the
people with the ideas but without the money – and there were investors
– the people with the money but without the ideas. We've discussed the
acute problem the entrepreneurs faced: getting money in their hands to
develop their ideas to make the world a better place, at least potentially.
But before you direct all your sympathy to the entrepreneurs, consider
the investors' position as well. Sure, it would have been nice for the
people with the money to indiscriminately fund every idea that came
along, thereby ensuring that the right ones would eventually rise to
the top.

---

[8] http://www.world-exchanges.org/statistics/annual/2009/equity-markets/
domestic-market-capitalization
[9] http://www.pngbd.com/finance/shares.php
[10] http://www.world-exchanges.org/statistics/key-market-figures

But there was a reason they didn't. There was a reason that they only tended to invest in the ideas of family members or others in their immediate circle who they felt they could trust. In today's world, what is the constant message we're given? It is to diversify our risk to every extent possible. For early shareholders and investors, with their limited supply of relatives, friends, or high-ranking connections, investment risk was virtually the opposite of diversified. If they did take the investment plunge, they could do it in only so many ways, after which they could do nothing but hope for the best. Returns were typically all or nothing – you paid for the ship, and then sat back to wait for it to come in with the spices and jewels the merchant had promised. As often as not, the ship sank (perhaps because it didn't have one of John Harrison's Longitude Prize-winning clocks on board), and the investor lost the lot, unless he'd taken on the additional cost of insurance, which was just as expensive and unpopular as it is today. In the case of Antonio in Shakespeare's *The Merchant of Venice*, losing one's lot meant coming perilously close to losing one's life.

Even if everything went according to plan, investors' hands were still tied as far as liquidity was concerned. Until the ship returned to port, there was no tangible way to realize one's investment. Traveling from say, Europe to the Far East and back, is no short trip, and to say investors had to wait a long time to see the fruits of their investments is a major understatement.

Banks, as powerful as they were, couldn't step in either. The design of the banking business model – slim equity supporting substantial amounts of debt – allows very little room for a bank to support risk in the loans it offers. Entrepreneurship is, after all, highly speculative, and the level of risk too high. A well-heeled private investor might feel like rolling the dice on an idea or two, but even the earliest financial institutions knew that it would take only a couple of risky investments to blow up before they were insolvent.

Entrepreneurs needed money to develop the ideas that might make our lives better. Investors needed a way to lessen their risk. Thanks to the Dutch East India Company, once the wheels of the capital market were set in motion, the mechanism for bringing money and ideas together only needed time to increase in both momentum and efficiency. As the markets both grew in size and expanded geographically, investors realized two benefits that had never existed before: more raw opportunities to invest, and a much bigger pool of potential ideas to invest in. These two new advantages meant they could finally diversify away many of the catastrophic risks that in previous centuries had been par for the course. A further advantage was the ability to maintain liquidity. Even though they had put money into something that wasn't guaranteed to succeed, at least it wasn't stuck on a ship at sea.

---

**Risk Diversification in a Nutshell**

Suppose you have a portfolio that pays off only when it rains. I, on the other hand, have a portfolio that pays off only when it doesn't. Someone approaches us both with a new investment opportunity that pays off only when it rains. If you add this new asset to your portfolio, your level of risk will remain the same – you'll still get a return only when it rains. But if I add it to my portfolio, I will get a return when it rains and also when it doesn't. The overall risk of my portfolio will have been reduced fundamentally. But you will have added even more eggs to the same basket.

---

## Making Balance of Lopsidedness

A further natural investment-promoting role played by the capital markets is to reduce what economists call information asymmetry. When a seller of something has more information about the thing being sold than the potential buyer, unevenness exists. The buyer is no less intelligent than the seller, but he is less informed. If he knows this, he will be much less likely to invest. The asymmetry, rather than stimulating the market, stifles it.

One cannot eliminate the asymmetry problem altogether, since there will always be knowledge disparity in some cases, but in the pre-modern era the magnitude of the problem was far greater than it is now. One modern response to information asymmetry is corporate disclosure rules. Another is improved accounting standards. Sometimes, businesses offset the potential information imbalance by simply clustering together. William Bernstein highlights how medieval markets, and in our day Hatton Garden in London – where around 300 jewelers occupy less than a block, serve to *"maximize the flow of pricing information to both buyers and sellers and increase the overall volume of commerce."*[11] Stock exchanges perform a similar function, not only by bringing buyers and sellers together, but also by acting as instruments for price discovery. By providing more dependable pricing, they increase the willingness of buyers and sellers to transact. As a result, the volume of capital available for getting us our stuff goes up.

## But Who Decides Which Projects See the Light of Day?

While the benefits of the capital market discussed above are of course important, the most important aspect of having capital markets is their influence over the *allocation* of capital. While the pharaoh may have been keen to have an enormous pyramid built to honor him in death, and certainly possessed the power to direct the lives of thousands of people squarely to this task, it is hardly clear that this is the allocation Egypt's people would have chosen for themselves had they had a say in things. Similarly, the incentives and prizes described above were driven typically by an individual or small group who held the power to recruit many people for a single objective without their having a choice in the matter. The key distinction between those prize-based or monarch-driven, allocations of resources and the capital market, is the matter of who is in control.

When the captain of a Dutch East India company ship is determining which products to bring back to Amsterdam from Indonesia, on what basis does he make this decision? Is the key factor his personal taste, or is it his ability to assess which products and features will be of highest value to consumers, which in turn drives the success of the company? Unlike pharaohs and their ilk, the CEO of the Dutch East India Company knows that in determining how to allocate the company's capital – people, ships, energy – he must understand and anticipate the interests, desires, and values of those who will eventually consume the things he brings home. The capital markets are able to orient the world's resources

---

[11] W. Bernstein, *The Birth of Plenty: How the Prosperity of the Modern World was Created*, (New York: McGraw-Hill, 2004), p. 134.

not around what a few rulers or committee members desire (or think the rest of us should desire) but instead around those products and services which we actually value.

The key here is the fact that the shareholders of the Dutch East India and its eventual customers were, by and large, part of the same group. If only one or the other, they would have had a difficult decision to make: I see myself first as a shareholder, and therefore try to realize the largest possible return on my investment, or I see myself first as a consumer, and therefore try to realize the highest quality and variety of products at the lowest prices. The two objectives are in conflict. If the shareholder doesn't benefit from lowering the Cost of Funding (COF) charged to the company, he would instead try to charge as high a rate as possible. But as a customer the solvency of the company is a secondary concern, and he might be happy to negotiate a higher quality and lower price until the solvency of the company is compromised.

However, those who were both shareholders *and* customers of the Dutch East India Company naturally found the balance through continuous re-pricing of the shares, such that the COF was low enough to enable the company to raise significant amounts of much needed capital to ensure it could access the resources needed to find, and bring back, the many products demanded by its customers. This needed to be done while the COF was still high enough to provide shareholders sufficient reason for keeping their money invested and thus support the ongoing operations of the company. This balancing act was achieved at a point where the Cost Of Funding (COF) was equal to the Opportunity Cost of Capital (OCC), a concept on which we will elaborate below.

## The Downside: Agency Costs

When potentially rewarding systems emerge, all sorts of characters typically show up to try to take advantage of them. The invention of the capital market is no exception, giving rise as it did to a brand new modern figure: the business manager. You might define the capital market as the transfer of ownership of the world's resources from the few to the many. And if everyone owns a certain something, it helps to have someone responsible for looking after it.

It didn't take long after the rapid climb of capital markets for a number of crafty individuals to step up and announce themselves as trustworthy types who would happily oversee and manage the assets owned by investors for a certain fee. And they were welcomed.

Simply put, the trade-off of diversification is that it's a lot harder to keep on top of your own portfolio. Having your money spread across different assets reduces your risk, but it makes things a lot harder to manage. This separation has thrown up what academics call the principal–agent problem. You – the principal – have money, so you hire a manager – the agent – to look after your investments. But all agents have their own interests. The manager you hire doesn't necessarily act to maximize your interests if his own are not aligned. As a result, your portfolio may not do as well as it otherwise might. The ensuing underperformance is referred to as agency cost.

Modern corporations make agency costs appallingly easy to understand. They include not only excessive pay packets but also the large expense accounts enjoyed by executives; corporate jets, lavishly appointed offices, the artwork that graces the walls of the executive suite visible only to the top brass but paid for with shareholders' money, and so on.

This sort of agency cost, where managers pay themselves too much or overconsume executive perks, is only the tip of the iceberg. The more pervasive the agency cost, the greater the room for rampant value destruction, and therefore the higher the likelihood of chronic underperformance due to lack of deliberate and focused effort from employees who aren't motivated to be as efficient and as productive as possible in delivering customer value.

Agency cost is the price we pay for the diversification benefits offered by the global capital market. However, we will try to convince you that being value driven and suffering agency costs need not co-exist.

---

**Agency Costs Can be Huge**

How big can agency costs be? In a word, big. In 1988, before the frenzied bidding began in what would become one of the most celebrated takeover deals of all time, food and tobacco giant RJR Nabisco boasted a market capitalization of about $13 billion. The buyout firm

KKR won a protracted bidding war with a final price just over $25 billion, borrowing over 90% of it. The deal was three times larger than any previous M&A transaction. Why? How could KKR pay so much when the market capitalization just weeks earlier was only about half this amount? Even more interesting for our purposes, how could RJR Nabisco's management have made a counter bid of $25.7 billion, even higher than the final selling price paid by KKR, taking the market cap to nearly double?

The answer is disconcertingly simple. Management merely submitted a business plan to the board that was completely different to the one that had been prepared prior to KKR making its colossal bid. It turned out that management didn't find KKR's assessment so far off after all, and they too could promise to just about double cash flows now that they saw the promise of an ownership cut. In other words, they were saying, *"As a manager, I'll manage the business to be worth $13 billion. But if you make me the owner, I'll manage it to be worth twice that."* The difference – $13 billion – was the agency cost.[12]

## In the Best Systems, No One is in Charge

In any company under the sun, the more employees are treated fairly and with respect, offered training and development opportunities, receive fair compensation for the value they contribute, and get time off to enjoy the fruits of their efforts, the more likely they will be to deliver value to the company – or least to want to do so. It should come as no surprise, therefore, that the companies that deliver the most value to their customers and endure the longest are often also the ones who pay the most attention to employee development, healthcare, work–life balance, and job satisfaction. No matter what industry you're in or

---

[12]The Board rejected management's bid, in part due to disgust over the conduct of Ross Johnson, the company's CEO. The decision had apparently been made that under no circumstances were they going to give the company to him.

the size of your company, every single person is involved in the value creation process.

Of course, if you really want to create value, it's vital to keep another thought in mind: even though everyone is involved, no one is in charge, or at least they shouldn't be. The capital market works because no one person or group of people decide how the money in it should be divvied up. Capital allocation is driven by some other criteria – but what? How do we make sure that, without thinking about it, capital markets direct funding to the ideas that will make a positive difference in our lives? To answer this question, let's discuss bacteria.

A single bacterium has no brain, yet collectively bacteria continue to thwart the efforts of medical researchers to kill them. James Surowiecki wrote about the wisdom of crowds, in which groups of people make better decisions than individuals because they possess far more information, even though they don't necessarily communicate that information directly to each other. Concepts that play on this idea, such as open-source innovation and crowdsourcing, which invite input from anyone and are ultimately self-policing, lie behind the success of Linux, Mozilla Firefox, and Wikipedia. The Internet, in fact, may be the perfect example of the ultimate democracy that is the collective brain. No one is in charge, so the only ideas that make it are those that truly bring value. All others die.

Capital markets, like the Internet, behave like a wise crowd – and the more investors involved, the better the result. There is no coordinated leadership, nobody trying to push things in one direction or the other, and, of paramount importance, no opinions to get in the way.

## The Relationship Problem

We are social creatures, and the act of forming relationships has benefited us since our distant ancestors learned the value of watching each other's backs, sharing food, and keeping an eye on each other's kids when the hunters were out trying to kill something for dinner.

In business, however, relationships aren't so useful. They cloud judgment and generate opinions. Opinions hinder the creation of value because they compromise objectivity and skew the collective democratic brain. Yes, relationships still benefit us in many ways, but there are times when we need to recognize them as more hindrances than enablers. Had the Dutch East India Company kept only to its immediate friends and relations instead of expanding its offering to others, it is highly unlikely it would have survived or thrived as long as it did – and for the purposes of this discussion, it wouldn't have unlocked such a powerful mechanism for putting money and ideas together and driving human innovation forward.

Just how much we tend to prefer those we already know was demonstrated by the acrimonious events involving the merger of Arcelor and Mittal Steel. Arcelor was a European steel giant formed by a merger between companies from France, Luxembourg, and Spain; Mittal Steel was founded by the Mittal family in India, later registered in Rotterdam. Its primary owner, Lakshmi Mittal, managed the company from his base in London.

Mittal made a bid to buy Arcelor in January 2006. The directors of Arcelor raised numerous objections, some having nothing at all to do with the business aspects of the deal. *The New York Times* reported that Arcelor CEO Guy Dollé dismissed Mittal as a *"company of Indians."*[13]

Arcelor announced in May that it was merging with Russian company Severstal. The CEO, Alexey Mordashov, was described by Arcelor chairman Joseph Kinsch as *"a true European,"*[14] and Dollé commented that *". . . because it is a friendly merger, the chance of success is very high."*[15] He went on to say, *"I have great confidence that the majority [of shareholders] will accept it."*[16]

Mittal, however, had already received the go-ahead from European regulators and launched its bid. Fighting back, Arcelor promised shareholders a higher dividend and to buy back shares at above market price. But shareholder pressure and an improved offer from Mittal – to the tune of €26.5 billion – eventually succeeded. The lesson here is that the capital

---

[13] http://www.nytimes.com/2006/06/04/opinion/04iht-edarcelor.1884078.html?_r=1
[14] Ibid.
[15] http://news.bbc.co.uk/1/hi/business/5018682.stm
[16] Ibid.

markets don't care if you are Indian, Russian, or something else. Nationality is something people see. The only thing the markets see is value.

The recent collapse of Ireland's economy offers a stark morality tale of what can happen when personal relationships are allowed to interfere with the allocation of capital. Anglo Irish, one of the country's largest banks, grew its business through aggressive, and careless, lending practices. As Michael Lewis writes, *"Anglo was able to shovel money out its door so quickly because it had turned banking into a family affair: if they liked the man they didn't bother to evaluate the project."*[17] Relationship-based lending meant no due diligence, and the bank's willingness to write checks for just about any amount requested by the borrower only made matters worse. Perhaps most shocking, lending officers were paid based on how many euros were lent – not on the quality of the loans, which you might imagine would be a more sensible guide.

To further illustrate this point, imagine a colleague or acquaintance of yours who was born and raised in Italy. He speaks English fluently but still carries a strong accent. Like any person, he has a number of good qualities and some less so.

What pops into your head whenever you see him? His personal attributes? Probably not. Because he looks Italian, dresses Italian, sounds Italian, you can't help yourself. All of us like to think we are open-minded, and can look past a person's nationality, gender, skin color, age, and so on, and judge them absent of prejudice. But we're human, and our personal biases are highly resistant to suppression. If we've had unfortunate run-ins with Italians in the past or traveled to Italy and didn't enjoy ourselves, we may harbor a bias against allocating rewards, energy, or capital to this person, independently of whether he deserves it or not. Just as harmful, we may consider Italy the most beautiful country in the world and everyone in it the salt of the earth, creating what social scientists call the "halo effect" – instead of a bias against allocating funds to this person, we convince ourselves that he is much more competent or trustworthy

---

[17] Michael Lewis, *Boomerang: Travels in the New Third World*, (New York: W.W. Norton & Co., 2011), p. 128.

than he actually is, and therefore highly deserving of funds. To a large extent, this is what happened in Ireland. Bank lending practices were little more than a family (and friends) affair, in which vast amounts of capital were directed to fellow clansmen without regard to whether or not their planned uses for it would create value or destroy it. We all now know the unfortunate result.

Karla Hoff of the World Bank and Arijit Sen of the Indian Statistical Institute have shown in specific ways how strong family ties can have a negative impact on economic development.[18] In most of Europe and North America, families tend to be defined as parents and children, with weaker ties to other relations. In many African and South Asian traditions, however, the radius of family is larger and stronger, extending to grandparents, aunts, uncles, great aunts, great uncles, cousins, second cousins, and so on. As McKinsey & Company's Eric Beinhocker points out, *"These extended family societies also have very strong norms on the sharing of economic wealth among family members. Richer family members are expected to help poorer family members. While the warm and fuzzy image of a large, extended family sharing with each other sounds very appealing and may have psychological and other benefits, in economic terms, it creates a basic problem."*[19]

Economists refer to this problem as moral hazard. Maybe you work hard at your job and manage over time to stash away some cash. Then your shiftless, good-for-nothing cousin comes to live with you and sponges off your generosity. After all, why should he work when he can count on your savings? Your financial support acts as a form of social insurance, allowing your cousin to be as lazy as he wants to be. The result is not only a free ride, incentive to freeload, and weaker incentive to work or save, but as Hoff and Sen point out, slower economic development at the country-wide level thanks to rampant nepotism in business and gov-

---

[18] K. Hoff and A. Sen, *A Simple Theory of the Extended Family System and Market Barriers to the Poor*, Sante Fe Institute Conference on Poverty Traps, working paper, July 20–22, 2001.

[19] Eric Beinhocker, *The Origin of Wealth*, (Boston: Harvard Business School Press, 2006), p. 434.

ernment, in which a handful of people work hard and make money and a large number of their friends and relations enjoy the benefits.

There is only one way to guarantee that capital allocation decisions remain objective: keep people out of them. The capital market achieves this goal because no one is in charge. The capital market is a cold, dispassionate broker, and we ought to be thankful. It couldn't care a whit about the color of your skin, your gender, your religious beliefs, your age or shoe size, the clothes you wear, the food you eat, the schools you went to, or who your parents are. It doesn't prefer Russians to Indians, or Italians to anyone else. It cares about one thing only: allocating capital to the products and services that deliver value to our lives.

To promote value creation in our own organizations, we must, like the capital market, remove personal biases and opinions from the equation. To be value-based in our decision-making means that our personal views, ethics, and beliefs have no influence whatsoever on how we allocate resources or make investment decisions. Ask yourself this: Who ultimately decides the value of the product or service you're trying to sell? Customers do. And do customers care about your personal beliefs and opinions, or do they care whether the thing you're selling will bring them happiness and make their lives better? The latter. And if those who determine the value of your business don't care about your opinions, why should you? You should care only about the same thing they care about, which is the same thing the capital market cares about: what creates value and what doesn't, plain and simple.

---

**Market Mimics**

Amazon and Google act like the markets because they are driven by data – no human bias is allowed to intervene in the drive for value creation. Amazon tells you constantly that *"people who bought what you've just bought also like this."* Google, through its extensive data mining, studies us and learns what we want, then its algorithms find out how to deliver it. It knows you've just done a search for rock climbing in Arizona, which might put you in a group of 2000 people who've done the same search in the past month. But you've also looked for Australian wine recently, which narrows the pool to about 200 – rock-climbing fans who also like Australian wine. It also knows you bought a motorcycle helmet not so long ago, which takes the number in your like-minded group to 20 or so, and after you've done a few more searches, Google knows you just about as well as you know yourself, or maybe better.

The advertising that appears when you do your next search is targeted specifically at you based on what Google knows you want. You might be presented with advertisements for rock climbing and motorbiking holidays in the wine regions of Australia. This is all done without a human making a judgment or offering an opinion. Just like capital markets naturally supporting some ideas and selecting out others, no one is in charge – and that's a good thing.

While individual biases and opinions are risky, collectively we humans can be relied upon to make value-creating decisions as a large group. Just as bacteria appear as intelligent as a super computer in large numbers, we can also look pretty impressive when we put our heads together, even if we don't realize we're doing it. Despite what we may think or want as individuals, via the market, we have demanded better, safer products with more features and more reliability for less money – think personal computers, cell phones, and any number of other items. We have made sure we are treated more fairly as employees, with safer working environments, proper support, opportunities for personal development, and higher pay.

The most recent manifestation of this phenomenon is our emerging collective demand that the managers of our resources treat our communities and our planet in a more responsible and considerate way. This is the concept of corporate social responsibility being driven by a market-like value stance, which is to say we realize collectively that destroying our planet would be bad for us.

And we won't take any excuses. Clothing giant GAP found out that even though it was unaware of the child labor being used in garment factories where its clothes were being manufactured in Cambodia, and even though the company wasn't breaking any laws, we as a collective consumer market didn't like it, and the company's share price suffered.

No individual person made it happen; with no brain, no heart, and no conscience, we jointly made GAP behave in a fairer, more moral way.

---

### The Power of Individuals

That the collective power of the market generates ongoing value doesn't mean individuals can't sometimes move mountains. When Cynthia Barlow's daughter was killed while riding her bicycle in London, she wanted answers.[20] It turned out her daughter had been struck by a cement truck that turned into her path because the driver hadn't noticed her. At the inquest, it became obvious that the driver should have seen Cynthia's daughter on the road, and that the accident could have been avoided. It also seemed evident that the company involved, known today as CEMEX, had no incentive to address the safety issues the inquest had raised.

Cynthia bought enough shares in CEMEX to be allowed to ask questions at the Annual General Meeting. She challenged the chairman and members of the board to explain *"what had happened, why it had happened and how they could stop it happening again. Everyone went quiet."*[21]

The Board listened, and now CEMEX truck drivers are trained using videos that highlight the dangers trucks pose to vulnerable users of the road. In addition, the trucks themselves have been fitted with extra indicators on the side and proximity sensors. The company even issued this awkward statement: *"Thanks to the adaptations we have made, CEMEX lorries have stopped killing people."*[22]

---

[20] Barlow later became the London chair of RoadPeace, a British advocacy group dedicated to improving road safety, holding dangerous drivers accountable through the criminal justice system and advocating for better rights for crash victims.

[21] C. Barlow (as told to Cole Moreton), "I bought shares in the company whose lorry killed my daughter", *FT Weekend Magazine*, August 7/8 2010, http://www.ft.com/cms/s/2/ca1fca5e-9f6c-11df-8732-00144feabdc0.html

[22] Ibid.

Cynthia's crusade, however, and others like it, are positive exceptions to the usual impediment to value that personal opinions create.
Just as a certain incentive may lead to the solving of the mystery of
longitude or a presidential exhortation may spur man to ultimately
reach the moon, individual voices can make a difference that leads to
increased value in our lives. But these are isolated incidents. Capital
markets, on the other hand, perform the same function day in and day
out around the world, not waiting for anyone to raise their voice but
constantly imposing the imperative of value on anyone who tests it.
If your company is publicly traded, the markets will tell you if you are
creating value or not.

We admit that there are times when the market gets it wrong, sometimes spectacularly so. Think of Enron, whose soaring stock price in the
1990s certainly suggested that the market thought Enron was creating
value. But Enron was lying – to its shareholders, to the markets, even to
itself – and as the news came out, the share price fell from a once glori-
ous $90 per share to less than a dollar. For the value-destroyers of this
world, a relentless and unforgiving reckoning awaits.

## Bacteria, Crowds, and the Markets

From their first appearance on the scene, capital markets were indiscrimi-
nate and all-inclusive – they encompassed every type of investor, providing a new, unbiased mechanism for bringing together the ideas that might
improve our lives and the money that could fund their development.
Because the markets had no close relatives, no high-status friends, and
no personal agenda, they were able to act in the blindly objective way
we as humans can't.

But while the capital markets have a naturally impartial structure, to
properly act in the interest of humanity, their triggers as to how to allocate capital must be driven by something, some kind of signal. Bacteria
respond to chemical signals that prompt them to synthesize molecules
or reproduce. People, often unconsciously, respond to a great variety of
signals, not always to their own benefit, of course.

The markets, too, respond to a specific signal: cash. When a given idea enters, the market responds to it – that is, "decides" whether that idea is value-creating or value-destroying – based on how much cash the idea is expected to produce, when, and according to what likelihood. The decision is generated by a silent collective intelligence with no one in charge and opinions placed neatly to the side, exactly where they ought to be.

But the markets don't just make a simple decision about how much an idea costs versus how much it might generate, or when, or based on what risk. They also make a decision about how that idea stacks up to others that might hypothetically take its place. We call this the Opportunity Cost of Capital – and it means everything.

# CHAPTER THREE

# THE OPPORTUNITY COST OF CAPITAL

*"Biology tells us that we are animals, and like all living things we exist only because we capture energy from our surroundings. When short of energy, we grow sluggish and die; when filled with it, we multiply and spread out. . . . we are unlike other animals only in the tools . . . that evolution gave us. Using these, we humans have imposed our wills on our environments in ways quite unlike other animals, capturing and organizing ever more energy, spreading villages, cities, states, and empires across the planet."*[1]

– Ian Morris, historian

The birth of the capital market gave entrepreneurs a new way to get funding so that they could build their inventions and make the world a better place while at the same time giving investors a way to diversify away risk. Investors no longer behaved like banks or the insular wealthy, allocating money only to people they knew. Entrepreneurs no longer had to depend on whether or not they could get the ear of the king. The dawn of capital markets meant that we had a much better chance of getting our stuff.

Another way of saying this is that the capital markets have gradually brought down the Cost Of Funding (COF). They have removed enough

---

[1]Ian Morris, *Why the West Rules – For Now*, (New York: Farrar, Straus and Giroux, 2010), p. 557.

risk to make it cheap enough and easy enough for entrepreneurs to get the dollars and resources they need to bring their brilliant ideas to fruition – at which point the market, in its inherent cooperative wisdom, decides whether to embrace the idea (if it recognizes it as value-creating) or reject it (if it sees it as value-destroying).

The basic initial function of capital markets then, has been to serve as a mechanism that could bring together the people with the money and the people with the ideas. But, as we said at the end of the last chapter, a truly high-functioning market makes an even more complex decision than merely whether an idea has value. It also decides whether the value of that idea is greater than that of a different idea that would use the same amount of time, money, or resources. Or, as we prefer to think of it, energy.

Investing is normally defined as the act of committing money or capital for the purpose of financial gain. We find this view too constraining. At its most basic level, investing is *the act of moving energy through time.*

As Ian Morris reminds us in the quote above, all life forms need energy to survive. Without energy, they revert to the particles they were before becoming life forms in the first place. *"Ashes to ashes, dust to dust,"* as the saying goes.

Consumption is the act of accessing and using energy now; investing, the act of denying oneself the immediate consumption of at least some energy in order to move it into the future for later consumption. Why do life forms invest rather than merely consume? Because they understand – all of them – that energy is essential to sustaining life, and that having energy today means nothing unless we also have energy tomorrow. For this reason, we invest; that is, we move energy through time.

In the past, humans invested by gathering and hiding nuts and berries or by hunting down beasts, drying, smoking, or salting the meat, and then storing it. For longer-term survival, humans made children, investing vast amounts of energy in the hope that those children would later return some of that energy to them in the twilight of their lives. No matter what form of energy they were keeping aside for future consumption, their main concern was that it would be pilfered by a neighboring tribe, or perhaps a pack of hyenas, or a group of mice, or some enterprising birds or bugs. This represented theft in the conventional way we think about it. But the more basic potential burglar of this stored energy was nature itself.

Even if other humans or animals didn't steal the food, it might simply rot. When this happens, we say the food has *"gone bad."* But the process is hardly so innocent. What we really mean is that the beneficial parts of the food have been appropriated by other organisms – microbes, mold, and so on. The sustaining energy has been seized upon by some other life form eager to consume or invest it just as we were when we obtained it in the first place. The same fate might befall our children – they might be taken by disease or infection from a minor cut. Nature wants the energy back and will be ruthless in reclaiming it.

It's hard to capture the energy we need for ourselves. The effort required to bring down a mastodon, or even to find the berries that will give us energy instead of poison us, is serious, and teaches us quickly that nature does not give up its energy without a struggle. If you doubt this, try to survive in the wilderness without the conveniences of modern living and see how far you get.

When we do capture energy, we would like to consume all of it, but a primitive instinct instructs us to invest some of it instead. This is in part an intelligent strategy to ensure we have energy for later and in part a defensive move. Investing means we not only get the energy we need to survive today, but we also keep it away from nature's manifold potential thieves for tomorrow. Investing is forever burdened with the problem that nature – that is, all other life forms – instinctively desires the same energy we try to hoard for ourselves. Here we see the most fundamental manifestation of opportunity cost, and the one that makes capital markets work.

To understand how opportunity cost drives the capital market, it is paramount that the term *capital* be properly understood. Capital is not money; it is stored energy. Allow us to step back and repeat ourselves a little in order to clarify the point.

To survive, we must invest energy, that is, we must figure out how to hoard enough of it to ensure our future existence. To obtain this life-sustaining energy, we must expend other energy to get it. *You have to give something to get something*, to quote another old saying. If you have to put out energy to get energy to be used for later, naturally the energy you use must be, at least according to your own perspective, a worthwhile expenditure when compared to the energy you acquire.

Every organism on the planet performs an instinctive calculation comparing the energy it must expend with the energy it might obtain for consumption or investment. This calculation is one of opportunity cost: the difference between what we are willing to put out versus what we stand to get back.

At least, every organism performs the calculation as best it can. We know that whenever we intend to claim some energy for our own use, nature will resist, and that this resistance will be represented by the competing instincts of all other life forms. But we can't ever measure this resistance precisely, so we rely on our insight and experience to provide us with what we hope are accurate guesses. When we guess right about opportunity cost, we thwart nature and its diverse array of potential energy competitors, at least temporarily. When we guess wrong, nature rules the day.

Put a shade in front of a tree and it will adjust the direction of its leaves to try to gain new access to the sun's energy. The tree knows that the alternative is death, so it makes an instinctive decision that the sun's nourishing rays provide greater benefit than the energy it must consume in order to access them.

Or imagine a deer that must consider leaving the protection of the forest to forage in the open meadow for food. The deer is only too aware of its increased visibility to predators should it choose to roam beyond the forest. It will expose itself to this heightened risk only if its instinctive calculation of opportunity cost tells it that there is a sufficient expected return on the energy it will consume. The energy offered to the deer by the life-giving grasses of the meadow must more than compensate for the increased risk of energy loss – in this case, total loss. No equation will guarantee the deer success. It must make one such computation after another, and it must hope, again and again, to be right.

## Fooling Nature

Battling every other organism for energy to invest is taxing even for us humans who reign comfortably at the top of the food chain. Several thousand years ago, our ancestors recognized a critical problem. Though they may not have phrased it the same way, the problem they sought to address was how to move energy through time without nature always trying to steal it back.

The answer they came up with was money. It's ironic that we tend to view money as the thing in our lives that carries ultimate value, whereas in fact the reason it is so perfect a foil for nature is that, in nature's view, it has no value at all. When our ancestors in their wisdom created the new system of trade, they cleverly decided to use as currency items whose energy was useless to other life forms. A gold coin, though it contains energy at the atomic level, has no consumption value to hyenas, mice, birds, bugs, mold, or any other organism.

Therefore, we can possess the gold indefinitely, since nothing else in nature will be using up its precious energy to come after it. At some future date when we need to obtain new energy, we can produce the gold coin and someone will provide us the energy we need in exchange. The one critical component of this system is "trust" – that is, given that the gold coin cannot sustain life intrinsically, it will only have value as currency if human beings trust that other humans will later give up energy in exchange for the gold coin.

We can think of the gold coin as a battery: a battery takes a charge of energy, stores it, and then releases the energy when needed.[2] The invention of money allowed humanity to trick nature with a technology that enabled us to move energy safely through time. The only exception, of course, came from within: other humans might try to steal the gold. This created the need for an institution that could safeguard our invest-ments – a topic we will discuss in detail later. The creation of money also enabled human beings to have a reason to change their approach to work. Without the ability to reliably move energy into the future, work effort was focused on short-term gain, and investment in work which would only pay off in the long term, was thought to have little value. However, with money and the ability to shift energy reliably into the more distant future, it increased the incentive to work on projects which paid off further into the future.

---

[2] James Rickards, *Currency Wars: The Making of the Next Global Crisis*, (New York: Penguin/Portfolio, 2011), p. 219.

## So How Do We Make the Right Guesses in the World of Business?

Are we suggesting that you stand up at your next board meeting and argue that capital should be defined as energy instead of funds or resources? No. But we are suggesting that heeding the comparison will allow you to make accurate decisions about the Opportunity Cost of Capital (OCC) when considering investments for your company.

At its most fundamental level, finance is concerned with transforming the idea of tricking nature into practical tools that allow us to allocate scarce resources, that is, bundles of reserve energy, such that the benefits of the energy we expect to gain in the future are worth more to us than what we have to give up in the present. In short, finance defines the OCC in terms of money, instead of energy. Whether you define the OCC based on money or energy, it involves the same objective: getting more out than what you put in.

But remember, we aren't just talking about whether you can invest 10 cents and expect to get 11 cents back. In business, the OCC for a given investment decision involves the comparison of what would be expected if the resources were to be invested in another idea of similar risk. Put another way, if you're presented with a brilliant proposal for a new product and asked to contribute a certain amount of your time, energy, and cash, say a million dollars, the opportunity cost of those resources is not the expected return on your million; it's what would be expected from them if these resources were invested in another opportunity of comparable risk. If the expected rate of return from the brilliant idea is 12% but the expected rate of return from the other investment is 15%, then investing in the first idea means losing 3% on your capital. The OCC is the benchmark of all benchmarks, because it represents the dividing line between value creation and value destruction.

The creation of capital markets did us the favor of driving down the COF by opening up the world of potential investment to the populace. The more functional the capital market, the closer the COF is to the OCC. In Chapter 2, we referred to the single most important element of the capital market: its ability to orient the allocation of the world's resources based upon the values and desires of consumers – *all* consumers, not just a few in positions of power.

We explained that it does this because shareholders are also consumers, so they naturally find a balance between their interests as shareholders – trying to earn a high return on their investments – and their interests as consumers – trying to obtain high-quality products and services with the features they value at the lowest possible price. The balance is achieved when the COF equates to the OCC, which is the happiness the resources could have delivered to the consumers had they been devoted to an alternative set of products and services instead.

In a poorly functioning capital market, the COF sits above or below the OCC, hindering overall value creation. If the COF is above the OCC, value-creating projects will be rejected again and again, since they will never generate a sufficient return. This was the state of the world before the Dutch had their revolutionary brainstorm in 1602. After that watershed moment, capital markets took root, gave rise to our ability to undertake value-creating projects which had been previously impossible, and ultimately led to the innovations that would drive the industrial and technological upheavals that brought us to where we are today.

This doesn't make our investment decisions automatically easy, of course. The capital market provided a mechanism; we need to produce value within that mechanism by making smart guesses. The crucial question then, is how to properly determine whether to pull the trigger on a given investment now that capital markets have so kindly set us up for potential success. To do that, you need to know how to value any proposed capital outlay; and to do that, in addition to being able to assess the happiness the project will deliver and the resulting cash you can expect to collect, you need to know how to get to the bottom of the OCC.

The opportunity cost of reading this book is the time and energy you're losing compared to the alternative of reading a different book, or watching television, or gardening, or napping, or doing any number of other activities instead. In the context of business, the OCC is what we give up by investing in one project or asset instead of another. It all seems simple enough. But we mustn't make the critical mistake of looking at the OCC from our own perspective. We need to remember that capital is energy, and therefore nature's perspective is the only one that matters.

Consider a business-unit manager within a division of a large company who is choosing between five potential projects. She might see the opportunity cost of allocating the capital to one of the five projects as the expected return her unit would miss out on from the best of the other four, and might therefore interpret the return she could have earned on that project as the OCC. But now think of her boss, the divisional manager, who is aware of not only her best alternative but also the opportunities available to the other managers in the division. To him, the expected return on her best opportunity is still well below the expected return on another project of similar risk that another manager elsewhere in the same division is considering. Thus, the business-unit manager's perception of opportunity cost is too constrained. From the division manager's perspective, her best opportunity isn't worth pursuing.

Now consider the group CEO, who has a broader view yet. Because she oversees every division in the company, she recognizes that the best return in this division is still not as good as the expected return from a proposed investment in another division. Having the luxury of seeing opportunity cost from a wider set, the group CEO is more likely to allocate capital more efficiently than the business-unit manager, whose scope is comparatively limited.

But wait – stopping here is as grave a mistake as would be accepting the perspective of the business-unit manager. Do we assume that the group CEO's perspective is adequately broad? Not quite. From nature's perspective, even if the CEO allocates the know-how, time, energy, and so on to the best of the available opportunities, it may well be that other companies in the same industry, or indeed in other industries, offer higher expected returns with the same risk. Exactly how and why are they able to offer a higher return with the same risk? If their plans to engage and employ people and other resources will result in a relatively higher level of happiness and will do so using resources more efficiently, then it is not unreasonable that expected cash flows will exceed those of the alternative company.

If that is the case, then nature – that is, the planet, our collective instinct for value – would prefer that these other companies get the resources. To properly embrace, and therefore assess, the OCC, we must give up our own perspective and pay attention only to that of nature. This is what the capital market silently asks us to do: allocate capital among a practically limitless array of investment opportunities in such a way that the greatest level of happiness is attained, as reflected by the risk-adjusted returns generated from the planet's scarce resources. If we can do that, we will continue to get our stuff, and we will do it more and more efficiently.

Are capital markets flawless? No. But for any "large" capital allocation system to work, it must deliver the goods. In other words, it has to channel capital in such a way that, on balance and over the long haul, ever more wealth is created. This is why capital markets have grown from almost nothing 400 years ago to a global behemoth. If a system fails to consistently select the best investment opportunities from the planet's perspective, it will make us poorer and we will naturally, at some point, do away with it altogether. This explains why alternative allocation systems have failed. Consider the Berlin Wall, and how the East German leadership adopted extreme measures to prevent the country's most valuable resources – its people – from exiting to the West. This system isn't quite held up as one we should all admire and seek to emulate. It was value-destroying, so we killed it.

In one of our presentations to a room full of executives, one, a European, remarked with considerable passion, *"This focus on value from the global perspective is very impersonal. I care about people, not value. I won't allocate jobs to China just because it creates more 'value.' I care about people, and I would rather focus on people and keep the jobs here rather than send them to China just because they will work more efficiently and deliver more happiness and generate higher returns on investment."*

After hearing the remark, one of the authors walked over to another of the participants, who happened to be Chinese, and said, *"Christophe doesn't think you are 'people.' He says he cares about 'people,' but apparently he only means 'people' who live where he lives and who eat, think, and speak the way he does. What do you think about this?"* The Chinese executive stated quietly, *"I think we are also people."*

Because of its naturally global perspective, the global capital market treats all people from all corners of the world with equal respect and consideration. Such behavior is nearly impossible for us humans.

## A Matter of Risk

We have said that to properly evaluate a given investment decision, one needs to properly evaluate the potential capital expenditure in the context of its opportunity cost. We have also asserted that this opportunity cost must be considered not from our own perspective but from nature's. Any time we try to obtain and use energy from nature, we need to acknowledge two factors: first, the amount of risk we're assuming when the expected future returns for the expended energy are certain, and second, the additional amount of risk we're assuming when those returns aren't so certain.

The first of these components refers to what is called the "risk-free rate," because it is applicable when the future return on the spent energy is certain. You might ask why the risk-free rate should be anything other than zero. If it is entirely certain that we're getting back our outlay of energy, what risk are we referring to?

The answer is this: we may know with certainty what the return will be, but that doesn't guarantee we'll be around to collect it. Remember that everything in nature, including the tree, the deer, and every other organism, is instinctively aware of its potential mortality. Even if the return is a sure thing, there is always the unfortunate possibility that one may die before getting a chance to enjoy it.

The notion of a truly risk-free investment isn't quite accurate, because even if we remove the risk regarding the level of repayment, we can never remove the risk regarding the ability of the investor to benefit from it. There is always an element of risk from the perspective of the party investing the time, money, or resources.

Do we have a way of determining the risk-free rate? We do. But first, let's discuss how not to do it. When many people learn finance for the first time, the teacher explains the concept of opportunity cost by asking: *"What could you have done with the money instead of putting it into the investment? You could have put it in the bank and earned the interest rate from the bank. Therefore, you passed up the opportunity to earn the bank's interest rate on the money. That is the OCC you are putting into the investment."*

While this is a perfectly apt explanation of the broad concept of opportunity cost, it is a deeply deceptive description of the OCC. The reason is that it is extremely personal, whereas determination of the OCC must occur from a much broader perspective than that of the individual investor or business manager.

The members of society have a choice about what to do with their money – invest it in real projects that deliver real products and services

to other members of society and, as a result of their willingness to pay
for these products and services, earn a return commensurate with the
value delivered; or, put it in the bank until such time as the first option
becomes a reality. When viewed correctly, we can see that the interest
rate paid by the bank is determined *by*, rather than being a determinant
*of*, the OCC. The OCC is first determined from the opportunities in the
"real" economy. If there are many excellent opportunities for investing
to earn a high expected return in real investments, then the bank rate
will have to be quite high to have any chance of attracting deposits from
people who would otherwise put their money into real investments.

While the interest rate paid by the bank may indeed be the opportunity cost of any *individual* person's decision to invest in a real project
or leave the money in the bank, it is not *the* OCC. Viewed correctly, the
bank's interest rate is itself determined by the OCC – as well as by many
other factors, such as the level of competition among banks to attract
deposits – rather than the other way around. To really understand the
OCC, we must look not to bank interest rates, but to the fundamental
drivers of those rates as reflected in the real economy itself.

The graph in Figure 3.1 below shows the trend in real GDP per capita
in the United States from 1870 to 2009. You'll notice that the rate of

Figure 3.1: Real GDP per-capita growth in the United States (shown on
a logarithmic scale)

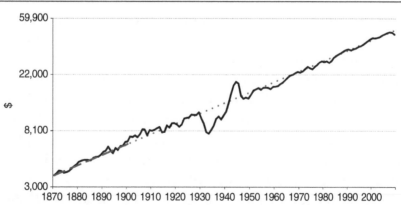

(Source: Antonio Fatás and Ilian Mihov, *The 4 I's of Economic Growth*,
INSEAD working paper, 2009)

growth fluctuates quite a bit over this period, which one can expect given two world wars, a great depression, numerous recessions, and staggering technological advances.

What is striking about the graph, however, is the consistency of the underlying trend. For the final 30 years of the nineteenth century, the average annualized growth in per capita real GDP was quite close to 2%. Since that period, despite all of the innovations and political upheavals that have occurred, the annualized growth rate has remained remarkably consistent. The implication of this trend is clear. An investor willing to place a long-term bet on the real economy in the US can rely on an annualized return of very close to 2%. This is the best estimate we have of the true risk-free rate. In order for banks to convince people to put their money into the bank, rather than into projects in the real world, the bank must offer interest of at least this amount (adjusted for riskiness, as we will discuss below).

Why are we using the US economy as the means for determining this figure? We are explaining that nature puts limits on the amount of return we can derive from our investments of energy, since nature could invest that energy in alternative pursuits and is in continuous competition with us to do so.

In what way is the evidence from the US economy helpful in ascertaining this number? Consider the individual entrepreneurs, investors, and workers who have comprised the US economy since 1870 – a truly remarkable sampling of humanity from across the planet, assembled in a common and predictable institutional environment, permitted to pursue happiness through a combination of production and consumption decisions nearly unhindered by any other force apart from nature.

What return would each of these individuals have "liked" to earn on their energy and efforts? Well, each would have liked a return of infinite percent, wouldn't they? Yet, while each sought to obtain the highest possible return on their efforts, the return which was earned, on average, was merely 2%. In other words, if you put enough people into an economy, with high levels of competition and the freedom to pursue individual economic goals, you will observe a rate of return that over time implies the level nature will tolerate. Sufficiently large numbers of human beings, acting without force or coercion from a common source, won't just look like nature, they will practically embody it. The US

Figure 3.2: Data for other developed countries (shown on a logarithmic scale)

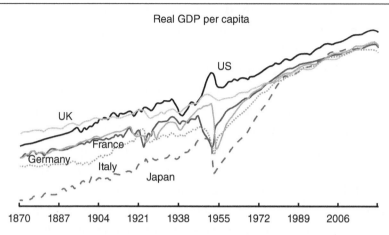

Real GDP per capita

(Source: Antonio Fatás and Ilian Mihov, *The 4 I's of Economic Growth*, INSEAD working paper, 2009)

economy offers us a large system with lots of people and a high degree of competition. Its long-run growth rates therefore represent our best estimate of the risk-free rate.

If you're willing to resist consuming energy today and instead choose to invest it (long-term) in the real US economy for tomorrow, you can be virtually certain of realizing a 2% annualized return, thanks to the giant laboratory that is the US economy.

But let's be clear. Under no circumstances should this number be thought of as US-specific. Remember, we are using the US economy as a proxy only because it happens to satisfy the right conditions. In Figure 3.2 we show data for other developed countries and they show a similar pattern once they also have the right conditions. The number is driven by nature, which means it applies to our human activities no matter where they're occurring, and we shouldn't be distracted by extraneous factors that make certain systems less reflective of high-functioning capital markets. If we were to perform the same experiment in say, Argentina, we would see a different trend. While the late nineteenth century would look similar for the two countries, from there onwards they would diverge, since early in the twentieth century Argentina's political leaders began moving their country in a very different direction from the US, resulting in less economic freedom than that enjoyed by Americans over the same period. This had a noticeably dampening effect on Argentina's economic growth. But remember, this means only that Argentina became a less accurate lab for the conditions that characterize nature at work, where the actors are free to behave without coercion or direction imposed from a single source other than nature's ruthless drive toward energy efficiency. Nature's inherent risk-free rate didn't change a bit.

When we attempt to interpret the risk-free rate in terms of money instead of energy, another variable comes into play. While energy remains energy, the value of money is not constant. For example, the price in US dollars of 2,000 well-balanced calories for an American consumer in 1870 was very different from the price in 2010.

There are two reasons for this change: productivity growth (i.e. value creation) and monetary inflation (i.e., the change in the value of the currency). The former leads to falling prices, while the latter causes prices to rise. If you are contemplating an investment in a money asset rather than a real asset, your expected return must compensate for any anticipated decline in the value of the money relative to the real asset. If a 2,000-calorie meal this year costs $10, and the inflation rate is 10%, meaning the same meal is expected to cost $11 next year, you're going to need eleven dollars one year in the future to have access to the same amount of energy you gave up this year. In other words, you'll need a 10% return in *money* terms just to break even in *energy* terms – plus the additional return in energy terms (namely 2%) to compensate for having denied yourself the consumption of the energy today to save it for tomorrow.

## The Premium

The second element of nature's OCC is the risk premium associated with any return that is less than certain. The risk-free rate refers to the risk of potentially not being around to collect on the return; the additional premium that we must incorporate has to do with the fact that the return itself is, in the first place, fraught with risk, never mind whether we'll survive long enough to collect on it.

Recall the deer searching for fodder. It has access to all the energy it needs if it enters the meadow, but it does so with the risk that it now becomes more likely to be seen by potential predators. In order to be willing to venture into the meadow, it needs an acceptable return for the added risk. This situation contrasts sharply with the discussion on the risk-free rate, because now there is some uncertainty about the payoff. Just like the deer, we all understand that higher risk carries a demand for higher return. If the deer knew exactly what return it would get in the form of life-sustaining grasses, it would still assume a 2% "risk-free" rate based on the chance that it might not survive long enough to realize the benefit. The additional question is how to measure the risk required to compensate for the added hazard of being extra visible in the meadow. As with the risk-free rate, we need some means by which we can observe nature at work.

The risk of a given investment is relevant only when viewed in the context of its impact on the riskiness of the overall portfolio. By portfolio, you can think in terms of wealth or energy. Either perspective will work, as long as we recall that wealth is useful only to the extent that it allows us to gain access to energy in the future.

Consider an investment that is risky in isolation (meaning its payoff is uncertain) but, when combined with other assets, actually reduces the risk of the portfolio. For example, imagine combining a business that sells sunglasses with a business that sells umbrellas. Both businesses may be risky in and of themselves, but when they are combined, the owners can be confident in their ability to feed their families come rain or shine. More importantly, the owners will be relatively indifferent to whether it is the umbrella business that generates the cash flow or the sunglasses business. This example demonstrates why the risk of an individual investment is not the relevant risk, but rather its contribution to the overall risk of the portfolio.

Since we know it's the riskiness of the overall portfolio that is relevant, we need to ask whose portfolio might help us determine the risk premium, and therefore the OCC. To answer this question, let's return to the example of the business-unit manager who had to choose from an opportunity set of five investments. From her perspective, the most pertinent portfolio would probably have been the existing investments already undertaken in her unit. She would have viewed any of the five investment opportunities in terms of how each of them might have contributed to the risk of her unit's overall portfolio.

However, the division manager or group CEO would view the riskiness differently, simply because they would assess the riskiness of each opportunity in terms of its contribution to a broader, and therefore different, portfolio. In fact, the project which may appear riskiest from the business-unit manager's perspective might well be viewed as the least risky according to the CEO. If the business-unit manager oversees the umbrella business and is considering a highly risky investment in that sector, it may seem excessively risky from her point of view. But the group CEO, responsible for not only umbrellas but also sunglasses, may know about a similar high-risk investment in the sunglasses division. The high-risk umbrella investment may offset the high-risk sunglasses investment, and therefore result in a lower overall risk for the business.

But recall that the CEO is in charge of only one company. The risk she perceives relative to the corporate portfolio may be very different from that perceived by someone able to consider the whole economy. Acquisitions are undertaken constantly based on the justification that they reduce risk. But this is true only from the perspective of the company in question. That company, and every other company, is part of the mechanism of the global capital markets, so the markets, in effect, already own them all. Therefore, combining two companies in a sub-portfolio accomplishes nothing in terms of risk reduction. Have we come to the right breadth of view, then? Is the capital market's perspective, as the owner of thousands of companies, sufficiently broad for our purposes? Not exactly – but it's pretty close.

As we've said, a well-functioning capital market is characterized by lots of participants, lots of transparency, and lots of competition. In such an environment, if there is an asset that is publicly traded and whose risks may be diversified away by being part of a portfolio with lots of other assets, yet is able to command a return for diversifiable risk, what would you expect a savvy investor to do? Buy this asset, and then buy the other assets needed to remove the diversifiable risk. In this way, the diversifiable risk will have been eliminated, but the investor will have captured an incremental return. Receiving a return for a risk that doesn't exist is a form of arbitrage. In a properly functioning capital market, such opportunities will be competed away. Because of this, the pricing, and therefore the expected return of every asset, will be based on the perspective of a fully diversified portfolio, even if no one investor holds this portfolio. It doesn't matter that most investors are neither savvy nor diversified; the market still acts as if they were. It merely requires enough investors to be savvy to arbitrage away the returns paid on any diversifiable risk.

What is included in this hypothetical portfolio? Quite simply, every asset on the planet – raw materials, real estate, human capital, minerals, water, the 1938 Superman comic that fetched over a million dollars in 2010, even your stamp collection. Of course, none of us humans can possibly hold such a portfolio, but there is one entity that does: nature. Thus the earlier point is re-emphasized. To understand risk and the role it plays in determining the OCC, we must assume nature's perspective.

Understanding that risk and return need to be assessed from the perspective of the fully diversified portfolio, which we can also refer to as "nature," the obvious question now emerges of how we estimate this number.

The answer is straightforward enough: find the portfolio which is as good a proxy as we can find for this fully diversified portfolio and which is traded in an active and transparent market. This ensures we are observing sufficiently large numbers of human beings acting in a highly competitive environment, so that we have a reasonable proxy for "nature." This was the same insight we used above to identify a reliable estimate for the risk-free rate.

Which market and which portfolio satisfy these requirements? We first propose the publicly traded US stock market, which "owns" or "rents" a vast portion of the US economic resources (physical capital, human capital, real estate, etc.) and which is traded in the most competitive, transparent and liquid market ever known to humanity (though it is still an imperfect estimator of the blue line, a point which cannot be overstated).

Using historical data from the US, we obtain different estimates depending upon which method is used for estimating the average (arithmetic or geometric mean), as well as for deciding upon which estimate to use for the risk-free rate (since the risk premium is that additional return above and beyond the risk-free rate that is paid on the fully diversified portfolio to compensate for the risk associated with payment), as well as the precise time period chosen. For example, using data compiled by Professor Aswath Damodaran[3] for the period 1928–2011, the geometric

---

[3] Aswath Damodaran, *Equity Risk Premiums (ERP): Determinants, Estimation and Implications*, 2012 Edition (updated March 2012), Stern School of Business, New York University, 2012, p. 27.

average annualized risk premium earned on the Standard & Poor's 500 portfolio of stocks over the Treasury Bond is estimated as 4.1%, while the arithmetic average annualized risk premium using the same data and time period is estimated as 5.79%. Recognizing that the choice between these is more a question of the appropriate holding period being used for both the investment and the generation of the cash flows being discounted, we might suggest that a better estimate of nature's portfolio is the lower of the two, since nature's perspective must be extremely long term and the geometric mean is a better indicator of the long-term holding period perspective.

On the other hand, since we are using the estimate to discount annual cash flows, it can also be argued that the arithmetic average annual return is a more appropriate measure. Recognizing both points, it may make sense to average the two, which provides an estimate of approximately 4.9%.

There are other financial economists who have used larger, or more recent, series of data to try to improve upon the estimates. For example, Goetzmann and Jorion use time series data on stocks and bonds from the period 1792 to 1925 and estimate the risk premium as 2.83% for the geometric mean and 2.76% for the arithmetic mean.[4]

However, we have argued that we want something as close to the "fully diversified portfolio" as possible, so why restrict ourselves to data from just the US? One reason is that it may not be the most fully diversified, but it has other important characteristics: high transparency, competitiveness, and liquidity to ensure that it is providing an estimate close to that of "nature." Nonetheless, researchers[5] have gathered data on the risk premium from other countries, a snapshot of which is shown in Table 3.1 below. For the combination of those 19 equity markets, from 1900–2011, the risk premium estimated with the geometric mean is 3.5%, and the arithmetic mean 4.8%.

As you can see, the estimates from the longer time series spanning back to 1792 for the US, as well as the data for other countries, provide lower estimates than the data from the US from 1928–2011. What should we do?

---

[4] Philippe Jorion and William N. Goetzmann, Global Stock Markets in the Twentieth Century, *Journal of Finance*, 1999, 54(3), pp. 953–980.

[5] Elroy Dimson, Paul Marsh and Mike Staunton, "Equity Premia Around the World", *London Business School* working paper, September 10, 2012.

Table 3.1: Historical equity premiums across different equity markets, 1900–2011

| | Stocks Minus Long-term Governments | | | |
| Country | Geometric Mean | Arithmetic Mean | Standard Error | Standard Deviation |
| --- | --- | --- | --- | --- |
| Australia | 5.60% | 7.50% | 1.90% | 19.90% |
| Belgium | 2.50% | 4.70% | 2.00% | 21.40% |
| Canada | 3.40% | 5.00% | 1.70% | 117.50% |
| Denmark | 1.60% | 3.10% | 1.60% | 17.20% |
| Finland | 5.20% | 8.90% | 2.90% | 30.40% |
| France | 3.00% | 5.30% | 2.20% | 22.90% |
| Germany | 5.10% | 8.50% | 2.70% | 28.50% |
| Ireland | 2.80% | 4.80% | 1.90% | 19.80% |
| Italy | 3.50% | 6.90% | 2.80% | 29.60% |
| Japan | 4.70% | 8.80% | 3.10% | 32.80% |
| Netherlands | 3.30% | 5.60% | 2.10% | 22.30% |
| New Zealand | 3.60% | 5.20% | 1.70% | 18.20% |
| Norway | 2.20% | 5.20% | 2.60% | 28.00% |
| South Africa | 5.30% | 7.10% | 1.80% | 19.50% |
| Spain | 2.10% | 4.10% | 2.00% | 20.80% |
| Sweden | 3.50% | 5.80% | 2.10% | 22.40% |
| Switzerland | 1.90% | 3.40% | 1.70% | 17.60% |
| UK | 3.60% | 5.00% | 1.60% | 17.20% |
| US | 4.10% | 6.20% | 1.90% | 20.50% |
| World-excl US | 3.50% | 4.70% | 1.50% | 15.60% |
| World | 3.50% | 4.80% | 1.50% | 15.60% |

Aswath Damodaran, *Equity Risk Premiums (ERP): Determinants, Estimation and Implications*, 2012 Edition (updated March 2012), Stern School of Business, New York University, 2012, pp. 30–31.

Unfortunately, there are additional arguments for whether the estimates calculated from such simple averages from the historical data are appropriate. Financial economists have proposed alternative methods for estimating the risk premium. These alternative methods include simply asking professionals, as well as estimating the risk premium as implied in the present-day pricing of equity relative to the risk-free rate as estimated by the promised return on present-day government bonds. Table 3.2 below presents these estimates in addition to the historical estimates. The average is 4.53%. While the pursuit of a single, "best" estimate for this number – the premium paid on the fully diversified portfolio over the risk-free rate – will continue for years to come, we shall stop our own pursuit at this point, and for our current purposes, accept an estimate of 4.5%.

Table 3.2: Estimating the risk premium from equity returns

| Approach Used | ERP | Additional Information |
|---|---|---|
| Survey: CFOs | 3.07% | Graham and Harvey survey of CFOs (2010); Average estimate.* |
| Survey: Global Fund Managers | 4.08% | Merrill Lynch (January 2012) survey of global managers |
| Historical US (arithmetic) | 5.79% | Arithmetic average of annual returns on stock minus returns on Treasury Bonds, 1928–2011 |
| Historical US (geometric) | 4.10% | Arithmetic average of annual returns on stock minus returns on Treasury Bonds, 1928–2011 |
| International Data | 3.50% | Average premium across 19 markets: Dimson, Marsh, and Staunton (2012) |
| Implied Premium | 6.01% | From S&P 500 – January 1, 2012 |
| Implied Premium Adjusted for T.Bond Rate and Term Structure | 3.50% | Using regression of implied premium on T.Bond rate |
| Default Spread Based Premium | 6.15% | Baa Default Spread * (ERP/ Default Spread average) |
| **Average:** | **4.53%** | |

Aswath Damodaran, *Equity Risk Premiums (ERP): Determinants, Estimation and Implications*, 2012 Edition (updated March 2012), Stern School of Business, New York University, 2012, p. 93.

*John R. Graham and Campbell R. Harvey, The Equity Risk Premium in 2010 (August 9, 2010). Available at SSRN: http://ssrn.com/abstract=1654026 or http://dx.doi.org/10.2139/ssrn.1654026

# Summary

Determining the OCC is a source of frustration for many policymakers, consumers, workers, shareholders, and myriad others. It is not something we can vote on or agree to set arbitrarily. If we agree to set it at 50%, we will be foiled when it becomes clear there are no projects that satisfy this minimum return. If we agree to set it at 0%, we will appear to reap many benefits from this new policy with many projects being undertaken, only to see that, within a decade or two, the willingness of people to continue to work to support these projects will diminish, and society will experience what will later be called the bursting of a bubble. This number is already determined, and we don't have the luxury of setting it where we'd like.

The OCC consists of (1) the risk-free rate, which reflects the situation where there is no risk of payment, and (2) the risk premium, which reflects the additional return required when a risk of payment exists. Each of these components has two sub-components.

For the risk-free rate, the two sub-components are (1) the "real risk-free rate," which reflects the amount which must be paid in energy terms whenever energy is invested, and (2) the inflation rate (referred to simply as "inflation") which reflects the decline in the purchasing power of the currency (i.e., the fact that it may take more units of the currency to buy the same amount of energy later than were required by that amount when the investment was originally made) and is needed to get the investor back to the starting position.

For the risk premium, the two sub-components are (1) the amount of non-diversifiable risk in the investment being considered (as any diversifiable risk cannot be justification for paying or receiving a higher return) and (2) the risk aversion of "nature" as estimated by the premium needed to be paid on the overall amount of risk in the fully diversified portfolio. From the data presented above, we have accepted the risk premium on the fully diversified portfolio as approximately 4.5%.

Putting this all together, we obtain an estimate of the global average OCC – for the moment ignoring inflation – of 2.0% (the risk-free rate) + 4.5% (the risk premium) = 6.5%. We recognize that this must be adjusted to reflect expected inflation when we are referring to investments in monetary terms; for example, if inflation is expected to be approximately 2.0%, then the nominal (i.e. that which reflects the additional return required to compensate for lost purchasing power of the currency due to inflation) OCC will be approximately 8.5%. For the remainder of the book, we will refer to the global average OCC as 6.5%, however the reader is reminded that this must be adjusted in any individual context to reflect the relative riskiness (in non-diversifiable terms) of the investment being considered, as well as expected inflation for the currency in question.

## The Mathematics of the Opportunity Cost of Capital

To reiterate, the OCC is the sum of two components: the "risk-free" rate, plus an adjustment for the added risk of an uncertain return. As we discussed, the risk-free rate consists of two parts, the real rate of interest and expected inflation. We can represent this relationship by the following equation:

$$R_f = (1 + \text{real rate of interest})(1 + \text{expected inflation rate}) - 1$$

To adjust for the risk premium, ideally we would observe the return on nature's portfolio and estimate the difference between this return and the risk-free rate we've already estimated as 2%. But we can't observe the return on nature's portfolio, so we need to find as close a proxy as possible and then allow as many people as possible to trade within this proxy in a competitive and transparent environment. Finally, we need to observe what aggregate return their combined activity generates.

Enter the capital markets again. Our proxy is an index of the world's major equity markets. It is not perfect, but more than adequate for the purpose at hand. The capitalized value of such an index runs to many trillions of dollars, large enough both in trading volume and number of investors to suggest that its risk/return tradeoffs are a reasonable representation of what we would expect to see if nature's portfolio were observable. If the risk/return profile of the index was significantly different from that offered by other asset classes within nature's portfolio – either more attractive or less – investment funds would flow in or out, causing the relative prices of assets within the portfolio to change such that the risk-adjusted returns from the index were no better or worse than those on offer from the complete portfolio.

We can then observe the average return on this proxy over a long period and accept it as an estimate of the expected (E) return (R) on nature's portfolio (N). We will denote this number as $E(R_N)$. To convert it into the risk premium, we need to obtain the OCC for a given asset and subtract from it the risk-free rate. In other words, we must perform the calculation:

$$E(R_N) - R_F.$$

Since this number provides the estimate of the risk premium in a well-diversified portfolio, we call it Nature's Risk Premium, or NRP. We will soon consider how different assets might have different amounts of such risk, and therefore will have higher or lower risk adjustments than the average captured in the NRP.

As we showed in the previous section, crunching decades of capital market data yields an estimated NRP of 4.5%. In other words, the best estimate of the premium nature demands for the risk represented in her portfolio is approximately 4.5% over the risk-free rate.

## How Do I Calculate the Opportunity Cost of Capital for the Investment My Company is Considering?

Let's consider an asset whose return moves, on average, by the same amount and in the same direction as that of nature's Fully Diversified Portfolio, or FDP. And let's define this relationship (between a given asset's return and the return on the FDP) by the Greek letter $\beta$. For this imaginary asset, $i$, we can then say that $\beta_i = 1.0$, since the asset's return is the same as the FDP's.

What return should this asset earn? Since it moves, in expectation, by the same amount and in the same direction as the market, we would expect it to earn the same rate of return, too. That is, we would predict $E(R_i)$, the expected return on this asset, to be equal to $E(R_N)$, the expected return on the FDP, nature's portfolio. For example, if the risk-free rate, $R_F$, were 4%, and Nature's Risk Premium, the NRP, were 4.5%, then $E(R_N)$, the expected return on the FDP, would be 8.5%, and $E(R_i)$, the expected return on our imaginary asset, would be 8.5% as well.

Now consider another asset, $i$, whose risk is on average half of the FDP's. In this case, since the amount of risk in $i$ is half the amount of that in nature's portfolio, $\beta_i = 0.5$. The opportunity cost of asset $i$ will likewise be less than that of the FDP. But by how much?

Recalling that the "risk-free" asset provides a return of $R_F$, $i$ must return this amount as a starting point, it will then pay an additional premium to reflect the fact that it has half the risk of nature's overall portfolio. This asset will have an expected return of $R_F$, the risk-free rate, plus $(0.5 \times NRP)$, half of NRP. Using the numbers from above (with the risk-free rate 4% and the NRP 4.5%), the expected return on this asset can be represented as:

$$E(R_i) = 4\% + (0.5 \times 4.5\%) = 6.25\%$$

We can see that if $\beta$, the relationship between a given asset's return and the return on the FDP, is less than 1.0, the risk is less than average. To put it another way, the expected rate of return is higher when $\beta$ is 1.0 than when it is 0.5.

Bringing this all together – combining the risk-free rate and the adjustment for risk, or risk premium – yields the Capital Asset Pricing Model (CAPM), the means by which we model capital market behavior to estimate the OCC for a company or individual investment project:

$$E(R_i) = R_F + (\beta_i \times NRP)$$

But capital markets do not determine the OCC, and the failure to recognize this fact is a source of great confusion. The market, by involving plenty of people with ideas and plenty of potential investors, determines the cost of *funding*, and this, as we have seen, can be very different from the cost of *capital* – the OCC. Remember that for most of human history, the COF has been far greater than the OCC, which is why so few value-creating ideas could be funded until recently. What capital markets offer is something else. By bringing millions of investors and capital users together in a trading platform with high levels of transparency and trillions of dollars at stake, the COF converges to the OCC. In effect, the CAPM allows us to coax nature's OCC from observable capital market behavior.[6]

---

[6]The asset pricing model we present here is more generalized than the one described in most finance textbooks as the Capital Asset Pricing Model. These discussions typically focus on the so-called Sharpe–Lintner CAPM. Unlike that model, which posits one non-diversifiable risk factor, our version allows for the possibility of multiple risk factors. Another difference is that our model relies on the arbitrage arguments described in this chapter (to explain, for example, why the cost of capital must be viewed from nature's perspective and not that of an individual investor). For these reasons, the CAPM described here is really closer in spirit to Arbitrage Pricing Theory. The one critical point on which the two models agree, however, is that investors can expect returns only on non-diversifiable risk.

## So Why *isn't* the Opportunity Cost of Capital Zero?

By now we hopefully agree that the OCC contains two parts: the risk-free rate (which we acknowledge is a slight misnomer), and the NRP, a premium to compensate for the uncertainty of the return. But here physics gets in the way of finance. When we explain this in the classroom, very astute students often point out that physics requires that the OCC equal zero. By this, they are referring to the first law of thermodynamics, or the law of conservation of energy, which insists that energy can neither be created nor destroyed, merely transformed. So how can the OCC be anything but zero?

The quick answer is that this law refers to a "closed" system, in which the total amount of energy never changes. Earth is far from a closed system, since we receive an abundance of new energy each day, in relatively constant quantities, from the sun. So while the law of conservation is true, it is not really applicable to Earth as a system.

Another way to say it is that, in the very long run, the risk-free rate and the NRP must both ultimately be zero in order to comply with all the others laws of physics that govern the rest of nature. But by "very long run," we mean billions of years. In the short run – meaning until the sun dies – it may appear as if we are defying the laws of physics, and in this context we can earn an OCC greater than zero. As long as the sun accommodates us by providing a constant source of new energy, continued value creation, and ever greater happiness for humanity, is a delightful possibility.

# CHAPTER FOUR

# THE EXPECTED FUTURE FREE CASH FLOWS

In Chapter 1, we argued that value creation from the customer's, that is, humanity's, point of view, is based on happiness, and that business inevitably was created as the means of delivering that happiness, in the form of products and services. We added that if value equals happiness for customers, then for the businesses that supply the happiness, the equation must be somewhat more complicated, since they must also stay afloat while furnishing us with the things that improve our lives.

To achieve their mission, businesses need to do two things: sell their products and services at prices consumers are willing to pay, and earn enough return on their capital investments to continue undertaking new ones.

To repeat our earlier statement, every decision a company makes either creates value or destroys it. To establish and maintain a constant value orientation, you need a way to define and calculate value in the first place. For customers, as we've said, value equals happiness. But for the businesses supplying this happiness, the equation is more complex, though in truth it involves only two parts. The second part of this equation is the concept we discussed in the previous chapter, the Opportunity Cost of Capital (OCC). The first part of the equation is cash – the cash you expect your products and services to generate. The full value definition is thus *The Expected Future Free Cash Flows Discounted at The Opportunity Cost of Capital*.

That's a mouthful, but every word in it is equally vital. In fact, the most important words in the equation are the two *"The"*s. We will explain why in a bit, in addition to explaining how to calculate whether an investment is value-creating or value-destroying. But first, let's talk about what value is not.

## It isn't Price

Oscar Wilde defined a cynic *as "A man who knows the price of everything and the value of nothing."* Warren Buffet said, *"Price is what you pay, value is what you get."* We say, when you get your stuff, it makes you happy, and there's no better definition of value than that. Your stuff has a value to you entirely unrelated to the price you paid for it. Or, more formally: price is the cash the consumer pays to obtain a product or service; value is the happiness he or she realizes from having obtained it.

What is the value of a can of Coca-Cola? Is it the amount of cash you paid for it? Or is it the satisfaction you get from quenching your thirst, or the pleasure you derive from its distinct taste? Say you paid a dollar for this can in New York. Is it worth more to you if you're in the middle of the desert?

What is the value of an old diamond ring? Is it the price a dealer will give you for it or the sentimental associations you carry from the imagined moment several decades past when your great-grandfather gave it to your great-grandmother?

What is the value of an innovative new surgical tool? Is it the cost of the raw materials used to make it, or the lives it has the potential to save?

Price is concrete and easy to define. Value is trickier – the value to you of anything you buy is not the same as the price you paid for it. Indeed, in any voluntary transaction, we can be certain of only two things: that the seller values the item at less than the price he sells it for, and that the buyer values the item at more than the price he pays. As a result, the value of the item will never be precisely equal to the price at which it was transacted.

A similar distinction should be made between share price and shareholder value. Share price is just what it sounds like: the price paid in the market for a company's shares. Shareholder *value*, properly understood, refers to the future cash flows that shareholders can expect to receive from the business. The two terms are often used interchangeably when in fact they have very different meanings.

In the interest of trying to raise a company's share price, for example, its managers may well make decisions that reduce the expected future cash flows, and therefore shareholder value. Even an executive as revered as Jack Welch has made such an error. In 2009, Welch told the *Financial Times, "On the face of it, shareholder value is the dumbest idea in the world. Shareholder value is a result, not a strategy."*[1] In using the term shareholder *value*, he actually meant share *price* – and in that respect, we agree that it is a result, and not the goal, of a properly run business.

But the parallel use of these terms leads to rampant confusion among the legions of managers who lack Jack Welch's talent for value creation. It is similarly common that managers confuse value with the outcomes of key performance indicators (KPIs). In other words, the results for certain KPIs, such as earnings per share or return on equity, are often taken to be synonymous with value creation, but just as share price is not value, neither is the outcome of a KPI. We'll have much more to say on this point later.

It's important to note that we aren't saying share price is unimportant. As we'll see later, share price is very important indeed – just not in the way most managers believe it to be.

## Getting at an Idea of Value

Let's return to the value equation: *The Expected Future Free Cash Flows Discounted at The Opportunity Cost of Capital.* We've already discussed the OCC, the most critical point being that we are concerned only with *the* OCC – that is, nature's – as opposed to what you, I, or anyone else perceives it to be.

It is vital that one regards expected future free cash flows the same way: as *the* expected future free cash flows, rather than those perceived, assumed, or determined by human bias or opinion. Companies willing and able to evaluate these two numbers – *the* OCC, and *the* expected

---

[1] F. Guerrera, "Welch condemns share price focus", *Financial Times*, March 12 2009, http://www.ft.com/cms/s/0/294ff1f2-0f27-11de-ba10-0000779fd2ac,dwp_uuid=c770f55e-0fac-11de-a8ae-0000779fd2ac.html

future free cash flows – and ignore the opinions of those who believe they know more than nature, will thrive. Companies who allow themselves to accommodate opinions, viewpoints, and hunches are doomed to failure.

Now that we have hopefully convinced you of the importance of the "the," let's proceed through the rest of the first part of the value equation – starting with the word "expected."

This word is indispensable because it describes all potential outcomes, each weighted by its probability of occurring. Imagine the following investment opportunity. You pay us to toss a coin, with the promise of receiving $50 if it lands on heads, $150 if it lands on tails. The coin is fair. We're going to do the experiment right now. What payment do you expect to receive?

When we pose this question to the participants in our executive programs, the most common responses are, as you might predict, $50 or $150. Virtually no one can (or is willing to) explain why they're choosing one of these two answers, but they're pretty comfortable that one of them is probably the correct one. Indeed, these answers would seem hard to argue with, at least on the surface. They are, after all, the only two possible outcomes.

But wait a moment. Instead of asking how much you would expect to win at this game, what if we instead asked how much you would be willing to pay to *play* this game? When we ask *this* question to our classes, things change in a hurry. Now no one knows at all what to guess.

Suppose you're willing to pay $110 to play. If the coin comes up tails, you get $150, a $40 profit. Are you happy? Would you do it again, paying the same price? In answer to the first question, of course you're happy – you just made $40. But if you continue to play, paying $110 for each flip, what will happen? In any given coin toss, you may get lucky, but after many coin tosses, you will almost certainly come out on the losing end.

More specifically, you will lose on average, $10 per coin toss – because while the *real* outcome for any given toss can be only $50 or $150, that is, in the real world, the *expected* outcome is what your brain was trying to get you to say when we first asked the question above: $100. In a large sample, for example a thousand coin flips, the average outcome will eventually converge to this expected number, and, just as the only opportunity cost that matters is that of nature, the only outcome that matters when trying to calculate value is the one nature expects, not the guesses you make.

In the classroom, we like to ask whether it was a good idea to pay $110 to play this game. The majority invariably answer "no," even if they gave the wrong answer to the other question about how much they would expect to win at the game. Their confidence in this response demonstrates the intrinsic understanding we all have of the importance of the concept of "expected" in value and value creation. It's easy to agree with the statement that paying more for something than you expect back is always a bad idea. It is, anyone would agree, obvious value destruction. The reason it still often happens to companies who understand the logic of the statement, is that they determine value based not on *the* expected but *their* expected, which is, we hasten to add, utterly irrelevant. All of the executives in the classroom expecting to receive $150 from the coin toss doesn't make it any more likely than all of them expecting to receive $50. No matter what, *the* expected outcome is $100, whether they agree with it or not. In case you aren't convinced, let's look at the probabilities and their predicted result, mathematically:

$$(\$50 \times 50\%) + (\$150 \times 50\%) = (\$25 + \$75) = \$100$$

This is an important concept to grasp. Value is the result of probabilities, in which one makes the best guess possible using all available information. Because value is probabilistic, it is totally disconnected to any forecasts or beliefs on the part of those calculating it. It is hard to separate ourselves from the mentality that the coin is going to land on either heads or tails, and thus pay out either $50 or $150. But this rejection of "personal perspective" is precisely what is required for effective projections of value. Value is a function of *the* expected cash flows.

It is not even based on the market's expectations. Say we created a competitive bidding scenario for the coin toss. Assuming enough participants, we would likely arrive at a price close to $100. It still wouldn't be *the* value, however it would represent a market consensus, which is a lot closer than the guesses of a bunch of executives in a classroom. We must look at value from a probabilistic perspective, not from the point of view of beliefs, whether yours, ours, or even the market's.

The role of probability in value was recognized by John Maynard Keynes nearly a century ago. Keynes asserted that an objective probability of some future event does exist. *"It is not,"* as he wrote, *"subject to human caprice. . . . A proposition is not probable because we think it so."*[2] Keynes illustrated this assertion with the story of the *SS Waratah*, a ship that disappeared off the South African coast in 1909. Re-insurance rates changed from hour to hour as bits of wreckage were found, and rumors spread that under similar circumstances a vessel had stayed afloat for several weeks before being discovered. *"Yet,"* as financial writer Peter Bernstein noted in his recounting of the episode, *"the probability that the* Waratah *had sunk remained constant even while the market's evaluation of that probability fluctuated."*[3]

Our point exactly. Individuals' perceptions may vary, and market prices for an asset may fluctuate, but none of this has anything to do with what the asset is *truly* worth, whether that asset is a coin toss, a share of stock, or a missing ship. Remember that value is a function of *the* expected cash flows, and these expectations are based on an underlying probability distribution. This distribution has nothing to do with what any person, or group of people (in other words, the market), may believe. Beliefs affect price, but they have no impact whatsoever on value.

Here's another way to look at it. Let's consider the distinction between value and price within the context of the development of modern physics.

Physicists have always been limited to the "reality" they were able to perceive. Their efforts, therefore, traditionally focused on trying to explain that reality with theories of underlying forces. As Einstein stated, *"What we call science has the sole purpose of determining what is,"*[4] with the obvious implication that there is, ultimately, an underlying determinism. However, the breakthrough thinking which led to the development of modern quantum physics was summarized by Niels Bohr: *"It is wrong to*

[2]John Maynard Keynes, *A Treatise on Probability*, (London: Macmillan, 1921), pp. 3–4.

[3]Peter L. Bernstein, *Against the Gods: The Remarkable Story of Risk*, (New York: John Wiley & Sons, 1996), pp. 225–226.

[4]Manjit Kumar, *Quantum: Einstein, Bohr, and the Great Debate about the Nature of Reality*, (New York: W. W. Norton & Company, 2010), p. 262.

*think that the task of physics is to find out how nature is. . . . Physics concerns what we can say about nature."* [5]

We mention this debate because associates of ours with physics backgrounds have noted parallels between the insights of quantum physics and our portrayal of value and price. Specifically, price represents an observation at a point in time, whereas value is the underlying truth – always. Depending upon the interpretation, this makes us sound either more like Einstein, arguing for a fundamental determinism (there is an underlying value for everything and it is true and determined and we are simply trying to figure out what it is), or more like Bohr (value does not exist except to the extent that it can be described in a probabilistic sense, and it only crystallizes into a "real" concept when a price is attributed to it).

We insist on the second interpretation. Value is in no way deterministic. It is determined by the expected future free cash flows associated with it, discounted at the OCC, and therefore exists in only a probabilistic way. When an exchange takes place, and a price must be arrived at and agreed upon, a new data point emerges for observation. But the equation remains the same.

Let us now consider the term "future," which demands its place in the value equation for two reasons. First, it emphasizes that the past is irrelevant where calculations of value are concerned. Putting it another way, sunk costs – cash that has already been paid or received – should never influence managerial decisions for the future. As the sayings go, *"Let bygones be bygones," "Don't cry over spilt milk," "Don't send good money after bad,"* and so on. Those who coined these adages seem to have had a good grasp of value.

The second reason we must include "future" in the value equation is quite simple. Anything that may impact on cash flows at any time between the point one makes a decision and any point after must be reflected in the value.

Few people would argue against including the word. But we often see confusion around its true meaning in managerial decision-making.

---

[5] Manjit Kumar, *Quantum: Einstein, Bohr, and the Great Debate about the Nature of Reality*, (New York: W. W. Norton & Company, 2010), p. 262.

In particular, two popular criticisms made against value-based management reveal a serious misunderstanding of what "future" means.

The first of these charges is that because capital markets are alleged to be short-term oriented, decisions taken with the intent to create shareholder value must lead to short-term decision-making. This criticism reflects both a failure to distinguish between shareholder value and share price, and also a lack of understanding of how capital markets work. It makes the mistake of imputing a short-term bias to the markets and misses the underlying truth: that individual market participants may simply be short-term in their thinking.

Recall our discussion regarding the OCC: while it is impossible for any one market participant to hold the Fully Diversified Portfolio (FDP), the market will nonetheless behave as if everybody does. Otherwise, arbitrage opportunities arise. Similarly, even if every market participant were short-term oriented, it doesn't follow that the market would follow suit. Indeed, if the market didn't factor in long-term expected cash flows, savvy investors could exploit arbitrage opportunities here, too.

To drive home the point, let us offer an example. Consider a hypothetical world in which every investor has only a five-minute investment horizon. In other words, the decision a given investor makes regarding which shares to buy and what price to pay, will be based on what she thinks she will receive in cash by selling those shares five-minutes later. The investor must perform two separate analyses in order to make this decision well: first, how much cash she expects to be generated during the five-minute holding period (you can think of it as a dividend payment), which requires a decision regarding the company's cash-generating capacity; and second, what price she expects the next investor will be willing to pay for the shares.

Now consider the perspective of a different investor who at the end of the five-minute period buys the shares from the first. In order to make his own investment decision for the coming five minutes, he must conduct an analysis that contains the same two questions: how much cash does he expect the company to generate during the holding period, and what does he think another investor will be willing to pay for his shares at the end of it?

Continuing this line of reasoning, it becomes evident that the only relevant factor is the cash-generating ability of the company in perpetuity.

No one investor need have a long-term horizon for the market price to reflect the infinitive one.

The second criticism of value-based management is that finance people in companies ignore the long-term impact of management decisions in their calculations. We don't disagree with this statement – but our point is that the underlying value of the decision does – indeed, must – reflect the long term, independent of whether the finance guys' calculations capture it or not.

## Free Cash Flow

We've moved through "the," "expected," and "future." Let's consider the next three words in this part of the equation, "free cash flow," as one entity. The free cash flow we refer to is the cash expected to be generated by the business and available for distribution to investors. To obtain this number, we follow the path provided by familiar accounting items – but recall that we are looking into the future, not the past, so while the concepts and account titles are similar to those one might see in a corporate financial statement, they are not identical.

Determining Free Cash Flow (FCF) is a somewhat straightforward exercise, though it does require a few calculations. We begin with the cash generated by the business from the products and services it sells, that is, its revenues. We then subtract any cash costs related to the units of product to be sold in that period, otherwise known as Cost of Goods Sold (COGS). We also deduct any cash overhead expenses related to running the business, which we call Selling, General, and Administrative (SG&A). From this we deduct Depreciation, but not the depreciation you would see on an income statement. Rather, it's the depreciation taken under tax law. This distinction is an important one because only the *tax* depreciation figure influences cash flows, and not the depreciation taken for financial statement purposes. From this we obtain Earnings before Interest and Tax (EBIT).

In COGS, SG&A, and Depreciation we have deducted the costs associated with running the business that will be paid to suppliers of raw materials, as well as service providers, employees, and all others from whom the company will obtain physical assets. However, we have not yet accounted for all of the costs payable to government. Some of these costs are included in COGS and SG&A, for example, employee-related health and unemployment insurance. But the most visible of these payments are the taxes calculated as a percentage of profit – in other words, income tax. In obtaining FCF, we wish to avoid any confusion associated with financing decisions and instead focus, to the extent possible, only on the cash generated by the business and available for repayment to all investors. Therefore, in defining FCF we do not subtract payments to investors. Most important, we do not subtract interest to be paid on debt. Similarly, any impact on taxes paid from how the company is financed is also something we will avoid at this stage. (We will address this point in more detail below.)

Therefore, to calculate FCF, we estimate the cash taxes to be paid by the company without taking into account the impact of financing decisions. That is, we estimate the cash taxes on the EBIT number, and denote this as the tax rate, $t$, times *EBIT*. The table below refers to it as *Taxes on EBIT* and it is estimated at the tax rate, in this case 25%, multiplied by *EBIT* (200 in this example), to obtain the figure in the table, 50. We refer to the resulting number as Net Operating Profit after Tax (NOPAT), calculated as $EBIT \times (1 - t)$.

But this is still not quite the net cash generated by the business. A few more adjustments are necessary. We deducted depreciation from revenues, but in fact the cash associated with the property, the plant, and the equipment which we are now depreciating, left the business at the moment the assets were bought. We deducted depreciation only to obtain the correct estimate of the cash taxes to be paid by the company. Therefore, we add back the depreciation deducted above to reflect the fact that the cash isn't leaving the company during the period in question.

We do, however, need to deduct any cash which is leaving the company this period for purchases of new assets that cannot be immediately expensed under tax law – those will already have shown up in COGS. These particular purchases are referred to as Capital Expenditures, or Capex.

Finally, depending on how we defined revenues, COGS, and SG&A, we may also need to adjust for the way those numbers were estimated. In normal accounting, revenues are booked before the cash from the sale is received, and the amount expected to be received from the sale is shown on the balance sheet as an Account Receivable until it is paid by the customer. Similarly, COGS or SG&A may not reflect the cash actually paid for the purchases of goods or services from suppliers, service providers, or employees because some of the items purchased may not have been sold yet, and thus remain on the balance sheet as inventory, or providers may have agreed to accept payment some time after providing the goods or services. The future payments for goods or services already received are shown on the balance sheet as Accounts Payable or Accrued Expenses. For this reason, revenues, COGS, and SG&A may not reflect actual cash received or paid. Adjustments will then be required.

We estimate these adjustments from the relevant items on the balance sheet by comparing what is booked with what is actually paid. The result is the change in the Working Capital Requirement (WCR), which equals the sum of the change in Accounts Receivable, Prepaid Expenses, and Inventories, minus the sum of the changes in Accounts Payable, Accrued Expenses, and, if relevant, Advances from Customers. We now arrive at a definition of FCF for this period that reflects all of the cash available for distribution to investors after all other constituents have been paid. And in so doing, we achieve a grasp of the entire value equation (see below Table 4.1: An illustration of free cash flow).

The next step is to determine the Net Present Value (NPV) of a given investment, which in turn will reveal which decisions will create value for your company and which will achieve the unfortunate opposite.

Table 4.1: An illustration of free cash flow

| | |
|---|---:|
| Revenues | 1000 |
| COGS | −600 |
| SG&A | −100 |
| Depreciation | −100 |
| EBIT | 200 |
| Taxes on EBIT | −50 |
| NOPAT | 150 |
| +Depreciation | 100 |
| −Increase WCR | −10 |
| −CapEx | −110 |
| Free Cash Flow | 130 |

## Net Present Value: Discounting the Expected Future Free Cash Flows at the Opportunity Cost of Capital

The additional complication arising from the fact that the choice of distribution (payment to debtholders as interest or principal payments, or as payments to equity), may impact the taxes payable to the government is not addressed here, though it is relevant to the value of the business, and we therefore pay attention to it below. Specifically, the tax impact of financing decisions is captured by estimating the cash impact of the tax shields and discounting the resulting tax savings to obtain the present value.

Since all of the expected free cash flows are in the future, if value is to have any practical decision-making relevance, we need a means of comparing the price we would have to pay today with the value of the cash flows expected in the future. In other words, we need to convert the future free cash flows to a present value equivalent. It is at this point that the two parts of the equation come together; we take the Expected Future Free Cash Flows and discount them at the Opportunity Cost of Capital, to finally arrive at our determination of value.

Consider a company that buys a computer for $2,000. The computer will be added to the accounting balance sheet as an asset with a "value" of $2,000. This, you won't be surprised to hear, is not the value we're after. What is the true value of this computer? We can be pretty sure it won't be $2,000, unless by some extraordinary coincidence.

The value of the computer comes from both the tangible and the intangible attributes associated with it. The former are easy to observe – the metal content of the components, the motherboard, and so forth – but they are the less important of the two. The real value of the computer comes from the intangible assets: what the computer will be used for, that is, the cash flows it is expected to generate by delivering happiness to customers. If it is to be used only for surfing the Internet over lunch, then the computer won't be of much value to the company, perhaps less than the price paid; but if it is used to find and capture new customers, or to design a more energy-efficient engine, its value will very likely be more than the price paid for it.

To be more specific, the computer's value depends entirely on whether the person who is going to use it knows how to turn it on, and then knows how to use the software needed to perform the tasks for which it has been acquired. Its value depends also on the reputation of the person and of the company with respect to other employees and customers, as well as perhaps to suppliers and other collaborators. And this value is subject to change, since while the tangible assets contained within it are constant, as it is hypothetically transferred from one user to another, or from one firm to another, the intangible assets associated with it go up and down.

Answering the question, *"What is the value of the computer?"* therefore means assessing what cash flows can be reasonably expected from the computer, which in turn depend on two factors: first, what the computer is used for and who will be using it; and second, the raw costs associated with buying the computer – that is, the capital required. NPV, thus can be defined as follows:

> The value of the expected cash flows, inflows, and outflows associated with a particular decision, discounted at the Opportunity Cost of Capital.

You may have already been exposed to this concept, and perhaps even have some experience applying it in your own business. Nevertheless, we have observed that, despite widespread acceptance of this idea, it is poorly applied in a great many organizations. So let's discuss the right way to think about NPV.

Any proper estimation of the expected future free cash flows for calculating the NPV must reflect three elements: first, all the outcomes that may arise in the future from a given investment decision; second, the probabilities for each outcome; and last, the cash-flow consequences for each outcome. If we return to the coin toss example, it seems simple enough. First, there are only two possible outcomes, heads or tails. Second, we know that there is a 50% probability for each. Finally, it is clear that one outcome yields $50 while the other yields $150. We arrive at the expected cash flow of $100 with little trouble.

But that is an undemanding example. The real world, of course, is far more complicated. The most obvious problem is that, unlike the coin toss example, the typical business decision has a virtually endless range of possible outcomes. Here it's critical to remember the importance of the two "the"s that appear in the value equation; allow us to remind you that *the* NPV of any decision must be distinguished from our ability to *estimate* it. Whatever level of difficulty we may have estimating a given project's value does not change the fact that it has one. Indeed, any business decision you make, or any resource allocation you undertake – launching a new product, entering a new market, hiring new staff, implementing an IT system, promoting and firing employees – will have an NPV, just as any decision you make in your regular life has positive or negative implications of some sort, whether you can estimate them or not.

To put it another way, *why* we need to rely on NPV for sound decision-making in business is completely separate from consideration of how to go about estimating it. Our challenge is first, for our managers to truly *understand* NPV and why it is the only relevant concept for driving decisions.

Second, we must have a process in place that both yields reliable estimates of NPV and enables these estimates to have decision-making utility. Never forget that for every decision and every allocation of resources in your business, there is an NPV, and that NPV is positive or negative – it will either serve to create value or destroy it. There is no middle ground. What is paramount to accept is that this fact is independent of your personal belief regarding the merits of the project. When we say that NPV is the only pertinent concept, we are attempting to guard against the tendency for managers to defend projects which they know in their hearts to be value-destroying but which they perceive to have some redeeming quality, like sustaining market share, hitting profit targets, or making employees happier.

Returning to our earlier example, how do we calculate NPV for our $2,000 computer? Suppose the riskiness of this investment is such that the OCC is 10%. Then suppose that buying the computer for $2,000 now will result in savings over the next five years of $600, $650, $700, $750, and $800 respectively. Is it worth buying, that is, is it positive NPV?

As Table 4.2 above shows, our free cash flows begin with the initial $2,000 investment, followed by the year-by-year cost savings. Now we must convert these cash flows to present value equivalents by discounting them at the OCC.

Table 4.2: Data for calculating the Net Present Value of an investment

|  | 2013 | 2014 | 2015 | 2016 | 2017 | 2018 |
|---|---|---|---|---|---|---|
| Cash Coming Back: |  | 600 | 650 | 700 | 750 | 800 |
| Investment Going In: | (2000) |  |  |  |  |  |
| Free Cash Flows | (2000) | 600 | 650 | 700 | 750 | 800 |
| Discounted Free Cash Flows | (2000) | 545 | 537 | 526 | 512 | 497 |
| Net Present Value | 618 |  |  |  |  |  |

The first positive cash flow, $600, is to be realized in 2014. Estimation of a present value of $600 to be received a year from now can be expressed as the question, *"What would you need to invest in 2013 in a project of similar risk in order to have $600 one year later?"* The answer will be the number we get when we add the return earned on the investment to the original number:

$$\text{Present Value } (\$600) \times (1 + \text{OCC}) = \$600$$
$$\text{So that, Present Value } (\$600) = \$600 \div (1 + \text{OCC}) = \$545$$

Thus, to get the present value equivalent of $600, we divided by (1 + OCC), and since we defined the OCC as 10% for this example, (1 + OCC) is 1.10.

To estimate the present value of $650 to be received in two years, we merely modify the question: *"What would I need to invest today in an asset of similar risk such that I would have $650 in two years?"* Our equation now changes to [$650 \div (1 + \text{OCC})^2$], which gives us $537. Continuing, the present value of $700 to be received in three years is $526, the present value of $750 to be received in four years is $512 and the present value of $800 to be received in five years is $497.

What about the overall NPV for the project? To arrive at this estimate, we must calculate the sum of these discounted free cash flows less the initial investment, or $545 + $537 + $526 + $512 + $497 − $2000, which yields an estimate of $617. Once we have this number, the decision is clear. The NPV is positive, so the idea is value-creating. The company should buy the computer, since, by investing $2,000 today, it will gain access to a future cash flow stream with a current value that is $617 higher. Put simply, undertaking this investment creates $617 in value.

What we have shown above is how to calculate NPV for a capital investment proposal. But remember, when we say the NPV is $617, we are talking about *the* NPV, which doesn't necessarily mean anyone will correctly estimate it.

The obvious difficulty is that in real life, we have the complication of *having* to estimate the Expected Future Free Cash Flows. Our ability to generate reasonable estimates is based on two key factors: first, a sound and proper valuation methodology that draws on the concepts introduced above, and second, an intimate knowledge of the business and every important aspect of it, including all major players in the company's value chain.

## Getting the Right Cash Flows: Modigliani and Miller

A constant challenge in understanding one's business well enough to generate reasonable forecasts of future free cash flows is to properly reflect the impact of financing. The contributions of two Nobel Prize-winning economists, Franco Modigliani and Merton Miller (popularly known as M&M), help us address this problem.

In the above discussion we defined free cash flows as those generated by the business without taking into account how the company's resources are financed. We briefly explained that one important route through which financing impacts these cash flows, is the company's tax payments. In particular, interest payments to debtholders are tax deductible in nearly all jurisdictions.

To illustrate the insights M&M have provided us, consider the following anecdote. One of us grew up on a chicken farm. His father financed the farm through a combination of debt and equity. Did the chickens care how much of the firm's financing was in the form of debt and how much was in the form of equity? Of course not. What about the other things needed to run a successful chicken farm? How much rain fell on the crops. How much the sun shone. How hot it became in summer. Fluctuations in the price of fuel for the tractors and combines in the fields. The price of eggs. Did any of these factors care how much of the farm's financing was in debt versus equity? No again.

So why would we expect that the free cash flows from the operations of the chicken farm would change depending on the capital structure? This is the essential insight M&M caused us to see: that, in order for the source of financing to impact the value of the company, it must impact either the OCC or the expected free cash flows. Whether it's in the form of debt or equity doesn't matter; if it has no effect on either of the two parts of the value equation, then it stands to reason that it won't have an effect on value.

Let us again split apart the two sides of the value equation and examine how each may be affected by financing. We know that the OCC is directly influenced by $\beta$, which, you'll recall, defines the relationship between the underlying company cash flows and the FDP. In order for financing to affect the OCC, it must have an effect on this fundamental relationship. By the same token, anything done to the company that is not correlated with the FDP cannot show up in $\beta$ and thus has no impact on the OCC.

Look at it this way. Increasing the debt level of your company may increase the riskiness of the debt, as well as the riskiness of the equity, but it says nothing about how the relationship between the riskiness of the underlying cash flows and the FDP might have changed. Until you can explain how the change in financing might have impacted that relationship, there is no reason to think that it has. Indeed, the requirement that the impact must be to that relationship, and cannot therefore be unique to the company, means that it will be extraordinarily difficult, if not impossible, for a financing decision to impact the OCC at all.

Yet it is commonly believed that financing choice *does* influence the OCC. We know that the OCC for any business is determined by the relationship between the riskiness of its free cash flows relative to the riskiness of the FDP. *How* the cash flows of the business are then divided among the different providers of finance does not, by itself, affect this relationship. We know this to be true because the relationship reflects only the non-diversifiable riskiness of the business, which we denote as $\beta_V$, whereas decisions about how to finance the business are taken by the firm's managers and are, by their very nature, uncorrelated with the FDP.

The resulting riskiness of the different financing components – debt (D) and equity (E), which we denote, respectively, as $\beta_D$ and $\beta_E$ – must

## Figure 4.1: Value and sources of finance: cause and effect for risk and the OCC

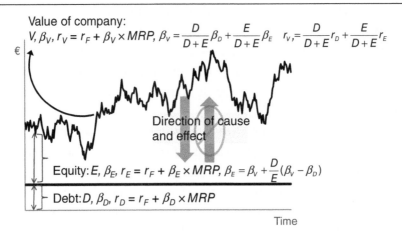

therefore simultaneously reflect (a) the riskiness of the overall business and (b) the allotment of the company's cash flows across its capital providers. This insight is represented by the following equation:

$$\beta_V = \frac{D}{D+E}\beta_D + \frac{E}{D+E}\beta_E$$

In Figure 4.1 above, the importance of M&M's insight can be seen by pointing out the direction of cause and effect. The fluctuating line in the upper part of the figure represents variations in the value of a company over time. The solid straight line represents the company's level of debt. The difference between the two, therefore, is equity. We can see that the riskiness in the movement of the value of the firm is not dependent on the line that separates debt and equity. Were we to move the debt up or down based on whatever capital structure policy struck our fancy, that policy would still have no impact on the value of the firm; it would influence only how much of that value belongs to debtholders and how much belongs to shareholders. The source of financing – in other words, the decision about how to distribute the firm's cash flows between debtholders and shareholders – does not determine the OCC for the firm's cash flows. It is the riskiness of the firm's cash flows, which, remember, is based on the risk of these cash flows relative to the fully diversified portfolio, that determines the OCC.

Since we know financing decisions cannot impact the OCC, if they are to have any impact on value, they must influence the other part of the overall value equation, the expected future free cash flows. And financing will almost certainly influence future free cash flows in numerous ways. For example, high levels of debt can influence the financial flexibility of the firm, the purchase decisions of customers, the terms of trade with suppliers, the likelihood of price wars and other competitive responses, and employee morale, to name but a few. Highly indebted firms may struggle to obtain financing for positive-NPV projects, miss out on managerial talent, lose customers who fear it will be unable to honor warranties and other after-sale commitments, and find it practically impossible to gain access to raw materials without paying cash up front. On the other side of the coin, firms with little or no debt may suffer from high agency costs. Managers, for instance, may become lax in the routine cash flow management of their business without the disciplining effects of debt.

All of these potential impacts will be felt in the expected future free cash flows. However, there is one critically important factor that requires special treatment, namely, the cash flow effects from the tax deductibility on interest payments.

What's different about the tax effects of interest deductibility is this: while the cash flow effects are real, they are not reflected in future free cash flows as we defined them above. The reason for this special treatment is complicated, but essentially it is a legacy of the M&M model that provides much of the conceptual foundation for modern corporate finance. We will touch on the issue in greater detail a little later in this chapter.

When we introduce income taxes, a portion of the firm's cash flows are distributed to the government, in addition to the cash flows that belong to investors. Note the difference between Figure 4.2 below and Figure 4.1 above. We now have two value lines instead of one: a solid line representing the value of cash flows before the payment of tax, and a dashed line representing the value of cash flows after the government has taken its cut.

Figure 4.2: Adding taxes to the picture

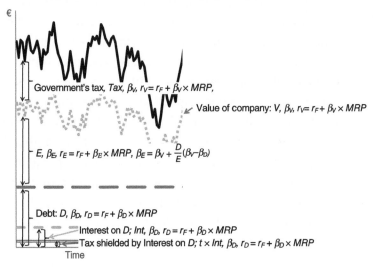

Capital Structure – Drivers of Value Impact

Before pressing on, we need to emphasize that the dashed value line does not include the tax effects of financing decisions. We thus refer to this line as the "unlevered" value of the company. However, changing the debt level (the gray large-dashed line) impacts the interest to be paid (the light gray shorter-dashed line), which then reduces the taxes to be paid (the double line). Mathematically, the taxes saved (the double line) equal the corporate income tax rate multiplied by the amount of interest to be paid. We must account for this additional cash flow, which is otherwise not reflected in our definition of free cash flow above nor in the value estimated based on that cash flow.

For a company with at least some debt, the value impact of that debt is felt in two ways. First, the cash flow effects from the tax shield afforded by the tax deductibility of interest ($V_{INTEREST\ TAX\ SHIELDS}$), plus all other effects captured in the free cash flows of the operations, as explained above ($V_{OPERATING\ FREE\ CASH\ FLOWS}$). These latter effects include the influence of debt levels on the behavior of customers, employees, and other relevant parties. The value of the levered company, $V_{LEVERED\ FIRM}$, is thus:

$$V_{LEVERED\ FIRM} = V_{OPERATING\ FREE\ CASH\ FLOWS} + V_{INTEREST\ TAX\ SHIELDS}$$

## What about Weighted-Average Cost of Capital?

When M&M developed their propositions 50 years ago, we didn't have high-speed personal computers, Excel, or even for that matter, calculators. Each stream of cash flows shown in the above expression had to be forecasted and discounted individually. Make no mistake, performing discounted cash flow calculations is not a quick process. As a consequence, M&M tried to simplify things by proposing an alternative formula. Under a set of restrictive assumptions, they showed that the present value of interest tax shields could be captured by altering the discount rate used to convert operating free cash flows to a present value. The modified discount rate they proposed is known as WACC, or Weighted-Average Cost of Capital. Their formula is as follows, where $r_D$ is the opportunity cost of debt and $r_E$ the opportunity cost of equity:

$$WACC = \frac{D}{D+E} r_D (1-t) + \frac{E}{D+E} r_E$$

With this formulation, we are able to take the operating free cash flows, discount them at the WACC, and obtain the value of the levered firm. There is no need to perform two separate calculations.

You may now be asking why we took so much space to explain how financing impacts value when a simple shortcut is available. The answer is that our objective in this book is to ensure that readers obtain a full and clear understanding of what value and value creation mean. The WACC shortcut has been the cause of enormous confusion and misunderstanding with respect to the concept of value generally and, more specifically, the OCC. Based on our observations of the thousands of people we have taught in our MBA and executive programs, there are two primary reasons for the confusion arising from the use of WACC.

Both problems arise from the very term itself. First, due to the exclusion of the word "opportunity," students of finance have been left to determine on their own whether the WACC refers to the weighted-average *cash* cost of capital or the weighted-average *opportunity* cost of capital. By failing to include an explicit reference to the opportunity cost concept as M&M intended, we see 50 years of finance students and practitioners interpreting WACC as the weighted-average *cash* cost of capital – in other words, the cost of funding.

As a by-product of this confusion, we also witness the majority of modern finance professionals asserting that if you can change your COF by negotiating with your banker, your WACC will in turn change, and therefore, so too the discount rate at which the value of the company is determined. This interpretation is wrong and inconsistent with a sound intuitive understanding of risk and return. Allow us to reiterate that the discount rate for the business is determined not by who provides the financing or the return they hope to receive, but by the riskiness of the cash flows, plain and simple.

The second problem with the term WACC comes from the notion of "weighted average." By stating that the discount rate for the free cash flows of the company is the weighted average of two or more numbers, the typical finance professional concludes that if any of those numbers change, the resulting weighted average should change, and therefore the WACC along with it. This leads to the false conclusion that the COF drives the discount rate, and that through this discount rate the source of financing therefore impacts the overall value of the firm. The particular irony of this error is that the key takeaway from M&M's work was that the impact of financing on value must come entirely through the cash flows – not through the discount rate.

Taken together, these two errors deriving from the use of the term WACC have had a profound effect. They have prevented the overwhelming majority of finance professionals from understanding finance.

# Chapter Five

# Blue Line Management

Having come to understand value conceptually – as the expected future free cash flows discounted at the Opportunity Cost of Capital (OCC) – we can take the next step, which entails asking a vital question: how do we manage for value creation in our day-to-day business? Recognizing the theoretical foundation of an idea hardly ensures that we will put it into action. A practical framework is needed.

To facilitate the application of value, we find it a necessary first step to distinguish between value as an objective and any other objective. In other words, if you aren't managing toward value, you are necessarily managing toward something else – and that something else, whatever it is, is, at least in our view, wrong. We call the path toward value the blue line, and the path toward everything else the red line. Companies that maintain a firm intention to create value are upholding what we call the Blue Line Imperative. Every decision they undertake is made based on one criterion and one only: whether that decision will create value or destroy it.

By contrast, companies who allow their decisions to be influenced by anything other than value creation are managing according to the red line – and that can lead to one conclusion only. It may take one year or 30, but red line management eventually spells doom for a company, no matter what business they are in. Blue line management, conversely, leads to growth, health, and sustainable competitive advantage.

## Value Creation and the Blue Line

As we discussed in Chapter 4, the value-driven company ultimately makes a rather simple decision regarding every investment possibility before them, even if the way to reach that moment of simplicity isn't so simple. The simple part, of course, comes down to a straightforward question: do we think this investment will create value or destroy it? As we've seen, such companies define this question in a very specific way – by calculating the decision's Net Present Value (NPV) as either positive or negative, via a specific equation, the expected future free cash flows discounted at the OCC.

To put it another way, blue line firms resolve to systematically invest only in projects where the expected value of cash coming in is greater than that going out – positive-NPV projects, in other words. If we wanted to say it in a slightly more complicated way, we would say that decisions expected to yield free cash flows with a present value in excess of the investment are value-enhancing, while all other projects are value-destroying. It sounds simple enough, but, as we've noted, these determinations of value depend entirely on the cash flows expected from the investment in a probabilistic sense, and not on the cash flow forecasts made by managers, which are by their very nature tainted by bias and opinion. At the risk of sounding repetitive, all investments have an intrinsic value, and that value exists quite independently of whatever you, we, or anyone else might believe it to be.

Confusing intrinsic value with personal estimates of value leads to the grave error we alluded to earlier: equating price with value. Price, again, is the outcome of a market mechanism, or negotiation, among two or more parties. For any item, from consumer goods to shares of stock, the buyer in a voluntary exchange privately assigns to it a value at least as high as the price for which she bought it, and the seller privately assigns to it a value no greater than the price for which he sold it. This price negotiated between the two parties will never, unless by extreme coincidence, be equal to the present value of the expected future free cash flows discounted at the OCC, and there is never any logical reason to assume that it would be.

Yet companies fall prey to these errors all the time. They allow perception and opinion to sway decisions, unintentionally ignoring value at their peril. Why do they do this? Because, as human nature dictates, it's hard to believe in something you can't see.

## Seeing the Blue Line

If we were to say to you that your company's blue line goes up every time it invests in a positive-NPV project, we don't imagine you would argue the point. Nor do we imagine you would rebut the assertion that the overarching goal of any truly value-driven enterprise is simply to raise the blue line as high as possible. Really we're just saying in the first place that positive-value decisions lead to positive results, and in the second place that the higher your proportion of positive-value decisions, the more successful your company will be.

But there is one very thorny problem pertaining to this concept which prevents many companies from putting it into practice: the blue line is invisible.

Why can't the blue line be seen? Because to know the intrinsic value of any potential investment decision, you would have to possess, and be able to accurately process, all available information about it, including all states of nature that might prevail in the future, not to mention the precise probability of each potential state of nature actually occurring, plus the cash flow consequences for each of these states. This isn't possible, of course, therefore the blue line, no matter how much you may believe in it, cannot be observed.

Price, on the other hand, couldn't be more observable. Why? Well, remember that price is generated by transactions, or by different people agreeing on forecasts of expected cash flows – which stands entirely separate from *the* expected cash flows that, in combination with the OCC, define intrinsic value. Forecasts may sometimes be reasonable estimates, but they can never be substitutes for the true probability distributions that underscore value.

But it is our natural desire to want to measure things – to make them tangible and real. Thus managers tend to focus on what they can directly observe, and it is these things that become their objectives. Quantifiable goals are comforting, even if potentially disastrous. You are no doubt familiar with these devils in disguise. They are known as Key Performance Indicators, or more commonly, KPIs.

Companies everywhere rely on a broad set of KPIs so that they can, they believe, accurately measure performance and keep their people focused and, for lack of a more elegant term, "incentivized." After all, how better to gauge the value of your company than by its share price? And what clearer way to motivate employees than by giving concrete, observable targets to hit? KPIs would indeed seem the ultimate carrots for inspiring value-generating performance. Unfortunately, this belief is most likely to have the opposite effect, and indeed has been responsible for staggering amounts of value destruction. The blue line is unobservable, yes. But it still represents the only goal for any value-driven organization.

## The Curse of the Red Line

Let's restate the Blue Line Imperative: an approach in which all decisions of consequence are made with the sole aim of creating value. This view stands in stark contrast to the more frequent practice of red line management, in which value creation may be the stated goal but the business is managed to deliver on specific indicator targets, independent of whether these efforts are value-creating or value-destroying. Here's a common example. The most *visible* indicator of what we will for the moment call "value" for a publicly traded company is share price. Indeed, value creation is often expressed erroneously as rising share price, as we have noted. But remember that share price is not value; it is the consensus of value by those participating in the market. Under no circumstances should it be mistaken for the real thing.

Consider the following graph in Figure 5.1, the red line (shown in gray) depicts the movement of a fictitious public company's stock price over time. We would expect the red line to fluctuate around the blue (shown in black), because, as deviations between the intrinsic value and share price get bigger, savvy investors will take appropriate action. If the shares appear seriously overpriced, they sell or even short; if the reverse is true, they buy.

The practical effect of this activity is the tethering of share price, an observable indicator of "value," to the blue line. The size of the tether separating the two lines – that is, the extent to which the shares may be mispriced – is a function of several factors, including the efficiency of

Figure 5.1: The blue line representing value and the red line
representing price

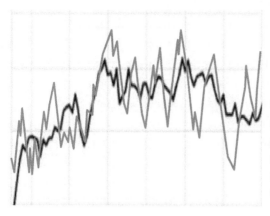

Price = Outcome of
a negotiated or
market process
clearing supply &
demand

**Value = The
expected future free
cash flows
discounted at the
opportunity cost of
capital**

the market, how widely held the shares are, the number of analysts tracking the company, and so on. Given that the shares of larger companies
are more widely held, and these shares are followed by legions of securities analysts, the extent of the mispricing is likely to be far less for a
global giant like Procter & Gamble than for, say, a small biotechnology
firm.

But here's the thing. Even if some investors were to know precisely
what the intrinsic value of a particular investment might be, the red line
and blue line would still diverge. The work of physicist and market trader
Doyne Farmer helps to explain why.

Farmer modeled a simple market with a single investor and a market
maker, and assumed that the investor knew exactly what the intrinsic
value of the stock was at all times and, in addition, that this intrinsic
value followed a random walk.[1] His expectation for the model was that
market price would track the random walk of true value. But he found
that this did not prove to be the case. Although market price tracked true
value roughly, the correlation was far from perfect.

Economist Eric Beinhocker, author of *The Origin of Wealth*, compares
this asymptotal relationship to adjusting the water temperature in a

---

[1] E. Beinhocker, op. cit., p. 394.

shower. We've all experienced this phenomenon in our travels. We turn on the shower in our hotel room, and the water is a bit too cold. We turn the handle slightly, but now it's too hot.

The key point to understand is that there will always be a time lag between our act of adjusting the temperature dial and the change in water temperature. Typically, we will overshoot, even when attempting to make our adjustments as fine as possible. We fiddle and correct until we finally get something close to the desired temperature.

Now imagine that the desired temperature were a random variable subject to constant fluctuation. That is, every time you get close to the target, it moves. The adjustment process must start all over again. Likewise, imagine our fictitious company's blue line, which is subject to a practically limitless array of macroeconomic, political, technological, social, cultural, and company-specific influences. Just as we tend to overshoot when trying to find the perfect shower temperature, the market will tend to overshoot similarly, continuously adjusting to the moving target that is the blue line. The red line, defined by observable, concrete indicators, can no more mirror the blue line than we could perfectly calibrate a shower whose desired temperature keeps shifting.

The problem is not with stock prices as such, nor with the issue of how price-efficient the capital markets may or may not be. It is, in a way, simpler than that. The problem is with the efforts of executives to manage price at the expense of value. While the red line cannot follow the blue line exactly, the two are of course inextricably related. Given the tether between the two, anything one does to detract from value adversely affects not only the blue line, but the red line in concert. Value destruction, in other words, causes share price to go down just as value creation causes it to increase.

Figure 5.2 below illustrates what happens to the red line as a function of changes in the blue line. Notice that the red line continues to fluctuate around the blue, but because the blue line is higher – a result of value having been created, the share price naturally goes up. If share price is to fluctuate around the blue line, isn't it better that it fluctuates around a higher level of value?

Firms that espouse the Blue Line Imperative focus on value creation as their overarching aim, with no exceptions. Red line-focused companies, or, as is often the case, red line-focused managers in companies that

Figure 5.2: The blue line drives the red line

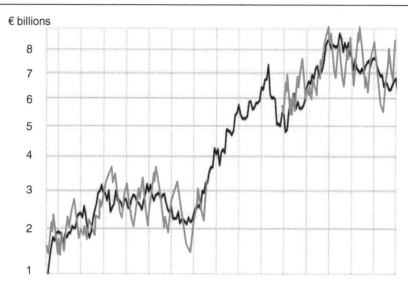

might otherwise be blue line, focus on performance indicators like share price, ignoring the blue line. Their stated goal may be value creation, but the practical reality is something else entirely.

Think of it this way. A blue line company never tries to "manage" its share price. It simply allows price to be determined by the capital market, knowing that if it makes positive-NPV decisions, one of the outcomes will likely be an increase in share price. The red and blue lines will never entirely coincide, but when viewed over a sufficiently long time horizon, they will converge to a high degree because, as mentioned, clever investors will try ceaselessly to spot and exploit mispriced securities. This means that the only way to be reasonably confident that one's share price will rise is to make decisions that raise the blue line. If a manager devotes time or other company resources to managing the red line, it is nearly certain that the diverted resources will cause the blue line to fall. The ironic result is that trying to manage the red line with the aim of raising share price will almost certainly cause it to fall.

Evidence that management in a given company is red line motivated will show up in a number of ways, both at the corporate level and business-unit levels. One common signal at the corporate level is the buying back of one's own shares with the aim of increasing Earnings per

Share (EPS); another is the massaging of accounting numbers to reduce earnings volatility. A buyback can be justified on several grounds, but doing it to increase EPS is certainly not one of them. This act is virtually always motivated by the misguided belief that share price is based on a fixed multiple of earnings. If this were the case, any action that increased EPS, even if it had no impact on the cash-generating capability of the firm, would cause share price to increase. Of course, the signal conveyed by an increase in EPS may be misinterpreted, but when the implied increase in future cash flows fails to materialize, share price is bound to fall.

Another common practice is to create hidden reserves of accounting profits through the over-provisioning of expenses and losses for warranties, bad debts, environmental cleanup, and restructuring. The reserves are then released in low-profit years down the road to artificially boost earnings. The effect of this practice is to create a stream of reported earnings that is far less volatile than that implied by the underlying operations of the business. Managers play this game for a number of reasons, but one of them is the false belief that income-smoothing can make their businesses appear less risky, thereby increasing share price. Through the lens of the red line, such games may make sense; through that of the blue line, they can lead only to failure.

## The Problem with Indicators

If used properly as a source of organizational learning – we'll get to that – KPIs can be highly useful. Unfortunately they can also be, and are often, used as red line tools. Negative value behavior is often motivated by the desire of senior managers to finesse share price, and KPIs offer a great way to do it.

As shown in Figure 5.3, value creation is achieved though actions and behaviors commonly known as value drivers – critical success factors that must be managed well if the business, any business, is to succeed. Examples of value drivers include employee motivation, customer and supplier relationships, and the configuration of equipment and machinery in factories.

But here we are met again with the problem of visibility versus invisibility. You might be able to gauge how well your relationship with your

Figure 5.3: The relationship between value drivers and indicators, and how they relate to value creation

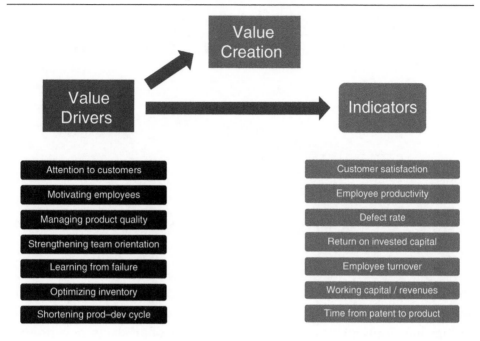

supplier is going, but you can't attribute to this value driver a specific, quantifiable impact. As it is not measurable, you instead attempt to capture it in a KPI, and this becomes a target to be met. The problem is that indicators are only the observable results of behaviors, and their outcomes encompass the influence of myriad other factors. There is no clean, direct causal relationship between a given business decision and its outcome as measured by a KPI. Between the time of the decision and the time of the result, a number of variables will have had an effect, many of them having little or nothing to do with the actions of management.

In other words, indicators are, at best, noisy measures of how well value drivers are being managed. They can be, as we have said, excellent tools for learning, but they must never be mistaken for value drivers, just as we must never confuse share price with intrinsic value.

Figure 5.4 illustrates the R&D process map for a large specialty chemicals company. For each box, there are behaviors that take place behind

## Figure 5.4: Drivers and indicators in a product development process

| What is the present value of the long-term cash impact of the effort? | How effectively do new products technology generate revenues &/or cost savings? | How effective is organization at implementing new technology products? | How effective is organization at converting new tech. to actual application? | How effective is organization at grasping and developing new ideas? | How effective is organization at identifying and acting on new ideas? | Is organization actively encouraging idea generation? |
|---|---|---|---|---|---|---|

**# ideas generated**

**# ideas introduced** X **% ideas acted upon**

**# ideas become actual technology** X **% conversion to patent/next step**

**# new products applications** X **% convert tech. to application**

**# implemented products/activities** X **% launch to client activity**

**Incremental cash impact** X **revenues cost savings per idea**

**Value creation** X **PV of expected future FCF**

**Time from idea generation to product launch / technology implementation**

the scenes to drive performance, and an indicator that summarizes observed performance for a given period. For example, converting technology into applications (enhanced functionality of products, new product features, and process innovation) is deemed to be a critical success factor, that is, a value driver. Because of its importance to the long-run success of the business, some effort is made to measure the performance of responsible managers on this dimension. However, here we run into the problem described above. As soon as you take a value driver and try to measure it by, for example, putting a dollar sign, point value, or percentage sign in front of it, the very nature of the value driver has been transformed. You have turned it into an indicator, and therefore something that it is, in fact, not.

We are not suggesting doing away with indicators. To the contrary, it is clear to us that any complex organization must be indicator-intensive if it is to judge how well key processes are being run or how they might be improved. The trouble occurs, as we so often see, when indicator outcomes become the focus of the business. Red line behavior spreads through all levels of the organization because employees are paid based on indicator outcomes, and it takes little time for them to discover that it is in their best interests to deliver on the indicator target, and thoughts of value be damned.

Returning to Figure 5.4, consider what happens when a manager is held accountable for converting technology into practical applications. A manager evaluated on the *percentage* of technologies converted may discourage or reject an idea that she believes is potentially groundbreaking because it has a high risk of failure. Meanwhile she embraces any project with a high probability of "success" even if she feels it will probably contribute marginal or negative value. She is gunning for those indicator targets, and meet them she will through this highly focused behavior. The problem, which is going to come back to bite the company, is that her focus is on the wrong thing. Her behavior, along with that of her fellow managers, will lead to a high observed value for this and other metrics, but at the very least, value-creating opportunities are regularly passed on. At worst, value is simply being destroyed, again and again.

The first problem leads to a second, and then a third. The initial problem is metrics being used as targets or incentives. The second is that when metrics are used as incentives, they no longer indicate what managers think they are indicating because they become altered by the efforts of decision-makers to manage them, an extra confound added to an already confounded measure of "value." The result is that even those employees trying to use KPIs properly – as tools of learning, diagnosing problems, or filling knowledge gaps – will obtain the wrong insights because the noisy proxies for value have become even noisier. Since everyone in the organization manages to deliver on KPIs, it becomes impossible to trust the indicators or to interpret them in a meaningful way. Managers will be unable to understand how changes in behavior impact either outcomes or value creation, and given the inherent difficulty in distinguishing positive-NPV projects from negative-NPV ones even in the best of circumstances, value creation becomes a practical impossibility because all of the numbers being used are, to varying degrees, lies. Used properly, indicators are instruments to promote learning and continuous improvement. But they work only if no one is manipulating or interfering with them.

The third problem is the most destructive. As employees are impelled to deliver on specific indicators, and they begin to manage and manipulate these indicators, they also realize that they don't have an aligned purpose. How could they? The company's mission statement contains the word "value" in several places, but how can this be so when the senior managers are evaluated strictly on the basis of one set of KPIs, like EPS, Economic Value Added, and Return on Equity – things usually associated with share price – while the employees themselves are evaluated on a different set which clearly support the managers' KPIs, but are nonetheless not the same. Furthermore, these lower-level managers have been told in no uncertain terms that their end-of-year bonuses are tied more or less directly to the indicator targets they are being asked to hit. If they are "incentivized" to do anything, it's only one thing: hit those targets at all costs.

Often, the message we hear coming from upper managers to lower managers is to deliver on the KPIs but not to compromise the long-term interests of the company in the process. Those receiving this message detect an immediate and obvious lack of integrity in the system, since they understand clearly that they cannot do both. What will most people do in this position? They will do whatever puts food on their family's table, that's what. If value is destroyed as a consequence, so be it – you need to look out for number one.

When we assert that this behavior is common, we are not speculating. To participants in our executive programs we always pose the following question: *"Have you ever been involved in or witnessed the willful destruction of value by someone in the interest of achieving a KPI target?"* Nearly everyone, including those in government and non-profits, are quick to concede that yes, they have observed such behavior on many occasions. Many sheepishly admit that they themselves have often committed the act – under-investing in a certain brand, or training to trim short-term costs, or playing revenue-recognition games to artificially boost sales growth. Why? To make the boss happy. More importantly, to not get fired.

Two deadly consequences arise. First, as middle- and lower-level managers recognize that they are destroying value as the only way to keep their careers on track, morale suffers. When morale goes, productivity declines along with it, since people begin to instinctively go through the motions. The company will have to start investing more to get employees to show up for work each morning and do a good job.

Worse, in such a system, employees come to understand that at least some of their colleagues who get promoted tend to be the biggest value-destroyers but the best at manipulating the system to their advantage. If morale was taking a hit before, it's facing a torpedo now. Morale sinks, productivity declines, and people who might have otherwise been eager to create value for the company, become resigned to hitting targets as a means of career survival. The red line has taken over.

Indeed, in organizations where the red line reigns supreme, even when senior managers insist that they are focused on value and encourage everyone else to do the same, middle managers recognize that they are being paid to deliver on indicators, with value placed squarely to the side. The employee we described above, who was trying to understand the disparity between the company's value-peppered mission statement and its indicator-driven system of compensation, now has a new quandary to tease apart. Senior management is trumpeting value as its focus, yet the remuneration scheme rewards value destruction, therefore senior management is either dishonest or confused. In either case, our dispirited employee does not see a common purpose pointing the organization forward. People are clearly not working toward the same purpose. If the people in a company aren't working toward the same purpose, what chance does that company have?

## A Blue Line Approach to Key Performance Indicators

If indicators should be neither carrots (to reward employees who hit targets) nor sticks (to punish employees who fail to deliver), what is their proper role in the value-driven company? The primary function of KPIs is to promote organizational learning.

Technological and scientific progress has given us products and services that accomplish things unthinkable a generation ago; but these advances have come at the price of ever-increasing complexity.[2] While it is virtually inarguable that products are more reliable, functional, and durable than ever, the business systems needed to deliver them have become devilishly complicated. A logical consequence of this complexity is ignorance and uncertainty. Systems are too complicated for managers to know all that they need to know to maximize value creation. Although companies may go to great lengths to design and document systems and processes, important pieces of the puzzle are frequently missed simply because the average puzzle has grown to such mammoth dimensions.

---

[2]For an interesting discussion on these issues, see Steven J. Spear, *Chasing the Rabbit*, (New York: McGraw-Hill, 2009), pp. 33–44.

We said earlier that to know the intrinsic value of any potential invest-ment decision, you would have to have all the information relevant to it and be aware of the probability of every possible state of nature in the future, as well as the corresponding expected cash flow implications. Even under the most basic circumstances, this is impossible. Today, given the levels of complexity within which products are conceived, designed, manufactured, and marketed, it is, shall we say, *really* impossible to possess such information. Knowledge gaps are more a certainty now than they have ever been.

An important part of value creation involves how we manage this unavoidable ignorance. In an uncertain, complex world, the successful business culture is one in which everyone is continuously learning. Of paramount importance, it is also a culture in which having the right answers is less important than asking the right questions. In other words, value creation demands experimentation. Blue line organizations don't just tolerate trial-and-error, they promote it. Only through continuous exploration and, yes, failure, can we gather new and relevant information to help us better understand the business and how to get more value from it. In a culture where people up and down the ladder are forced to obsess over hitting indicator targets that represent false value, the blue line cannot possibly take root.

Indicator targets *are* important, but not as performance motivators. When set honestly – that is, absent of the game playing so common in budgeting these days – targets become hypotheses about how business systems are supposed to work. As such, they reflect expectations based on imperfect knowledge, and that is exactly what we want them to reflect, since that is what will lead to continuous learning and improvement. Properly calibrated KPIs will reveal the extent to which our understand-ing of the system was incorrect or incomplete – a confirmation or repu-diation of the hypothesis. In this way indicators can act as highly useful warning signals, the proverbial canary in the mine shaft, since discrepan-cies between the targets (the hypothesis) and the outcomes (the result of the experiment) can reveal critical problems as they emerge, allowing us to plug knowledge gaps and resolve problems faster. Or, they may validate the hypothesis and provide important evidence that the invest-ment seems a positive one.

But this process can work only if we allow it to. The measurement of outcomes must be unbiased, and fully uncontaminated by efforts to steer or massage the indicators. KPIs always have something valuable to say, but the value must be allowed to emerge pristine. If the business is managed *according* to the KPIs, then potentially indispensable learning will routinely be suppressed, hidden, or ignored.

It is easy to set a target, reward those who hit it, and withhold reward from those who fall short. Companies that organize themselves like this, may say they operate in a way that is very clear and obvious to everyone within the walls. You know what you have to deliver on, which makes it easier to do the job well. But punishing people for failing to deliver on KPI targets is like punishing a scientist for conducting a lab experiment whose hypothesis is not confirmed. In a red line culture, that scientist has failed his experiment. In a blue line culture, he has learned something of value. And isn't business just a never-ending series of experiments? If this is the case, how can we punish people for the experiments that don't pan out?

It's no revelation to say that critical learning often occurs through failure. However, even knowing this, many companies default to the red line because it seems simpler and more comfortable – with more visible things to measure – and therefore they encourage extraordinary amounts of time and energy to be spent hiding "failure." To turn it around, they can begin to do something different: use those same resources to encourage learning, and thus create value.

## Getting Indicators and Value Drivers Straight

Here's a great example of blue line management we recently witnessed. A group of senior managers from GE Capital visited one of their European investments and interviewed the head of sales. They started out by asking him a series of questions. They included inquiries like the following:

- *"How many calls do your relationship managers make each day?"*
- *"What percentage of these calls gets through to the decision-maker?"*
- *"How many decision-makers agree to a meeting?"*
- *"How many meetings do they book?"*

- *"How many deals close?"*
- *"What is the average deal size?"*

The head of sales, an MBA graduate recruited from Goldman Sachs, couldn't answer any of these questions with precision. When asked how he managed people, he immediately relaxed, clearly now on safer ground. He described the series of awards given for best deals of the week, month, quarter, and year, ranging from a bottle of wine to a weekend away at a nice hotel.

Curious, the GE Capital team pointed out that this answer had to do with what people were paid, not how they were *managed*. Their next question was, *"What actually, do you do? You don't manage anyone, you don't know the processes for your part of the business, and you don't know the results, so what exactly are you being paid for?"* The head of sales was taken to GE Capital's offices in London to learn about management.

Before this trip to London, he had held a monthly meeting with his relationship managers in which two very clear goals were accomplished: first, a discussion on what was achieved in the previous month, and second, an agreement on a target for the month ahead. At the following month's meeting, they would discuss the target they had set, and usually why it had been missed.

The relationship managers who worked under the head of sales would begin each month fired up, determined to crack the code and finally hit the target, but then they would become progressively demoralized as the weeks went on and it became clear once again that they were going to fall short of their set goal. They also became nervous. The company was young, so they didn't have a well-known brand name to back them up. They spent much of their time preparing beautiful presentations and elaborate spreadsheets for the meetings that weren't actually scheduled but which they were sure they would eventually book. When the end-of-month session with the head of sales arrived, it would be an inspiring confessional in which they apologized for failing and promised, again, to do better next month.

When the head of sales returned from his visit to GE Capital, he did something new, instituting a daily "power hour" that started at 4 o'clock every afternoon. In each country office, all relationship managers were now asked to sit around a table with their phones and a list of 50 companies that they were responsible for developing. They sat where they could see each other – there was no place to hide – and began to make calls the moment the meeting began. At 5 p.m., they stopped.

From the daily power hour, the newly reformed head of sales started to do something he had never done before: actually gather information about what was driving value in the business. He was able to determine the average number of calls a relationship manager could make in an hour, how often they were able to get through to the decision-makers, and how many meetings got booked as a result. Everything was documented and displayed, and the relationship managers hated every minute of it. No one likes cold-calling in the first place, and they had been doing everything possible to avoid it. Being forced to do it in front of each other, and then having the results up for everyone to see, was even worse.

But learning is learning, and it is valuable no matter what the situation. The head of sales began to discover all sorts of things about how his business was being run and how it could be improved. The conversion rates, for example, made blindingly obvious which relationship managers were better at getting through to the decision-makers and which were not. With this information, the head of sales was able to have his less-effective relationship managers model their actions on those of the more successful ones. One simple lesson from those at the higher end, for instance, was that persistence paid off – they worked hard at getting to know the gatekeepers, called back regularly, and eventually made inroads. As a result, their conversion rates were higher. It probably seems foolishly obvious when you read it, but without having created the daily forum for learning, the head of sales would not have had any way of knowing this, or of doing anything about it.

And that was just the tip of the iceberg. Information about the conversion rates also helped demonstrate that spending more than five minutes on a given call was essentially useless, since beyond that point meetings were almost never secured. They were, at that point, explaining too much, showing all their cards while on the phone, and therefore letting the prospect make a quick decision instead of finding the opportunity to explore their needs and tailor a solution. After five minutes, it was time to push for the meeting and get off the phone. Having that kind of data in the arsenal led to higher conversion rates for everyone.

Freshly convinced of the Blue Line Imperative, the head of sales also started to use the numbers to conduct experiments. He explored, for instance, the question of whether it mattered what the relationship managers wore and found that conversion numbers were better for those who wore suits than those who dressed more casually. No, they were not interacting with the potential customers face to face, but data is data.

He also investigated whether there was an effect based on the way the company was referred to. For one week, he instructed the relationship managers to always add "dot com" to the company name. For the week following, he instructed them to drop it. Though everyone in the company assumed the addition of "dot com" would make them sound cooler and more intriguing to prospects, the numbers (e.g., client decision-makers agreeing to a meeting, the number of meetings booked, the number of deals that close, etc.) were dramatically lower in the first week, and so the addition was dropped like a hot potato.

He didn't stop there. His next step was to attend the meetings with prospective clients alongside the relationship managers, in which he began to collect similar information to the power hour. This time he observed which presentations seemed to work and which didn't, which relationship managers were better at interacting live, which were able to follow the process through to signing, and so on.

Finally, he discarded the old system of incentives and replaced it with one that offered each relationship manager a share of the total profit. After all, it was pretty clear who was adding value by making calls, getting meetings, and closing deals, and who was essentially freeloading. Incentives could be based on successes rather than hitting the target for the number of attempts.

An energetic buzz took hold in the office. The relationship managers began to feel more like a team, and even began to check the numbers to see who was doing well and who should ask for help if they were stumbling. People started to feel good about themselves because they were doing their jobs, as opposed to putting off unappealing tasks all month and then dreading the awkward meeting in which they would have to explain why they hadn't hit their KPI targets. With everyone working together to increase value and therefore their own share of the pie, synergy and transparency went up. Any one of them could succeed only if they all succeeded, so they had no choice but to work hard together, and it felt good.

In asking the head of sales about the number of calls and how successful they were, the GE Capital team well understood that the former was merely an indicator, not a representation, of value. How many calls you make doesn't matter much at the end of the day; what matters is how many of those calls become sales. It's the conversion process that constitutes the value driver, not the making of the call itself. The head of sales, properly edified regarding the blue line, had realized that he should use the numbers to learn – to signal how to improve processes. He thus learned how to use indicators properly: as a means of understanding drivers. If you want to be a blue line manager, stop managing the indicators. Use them to understand where processes can be improved and extra value created.

## When Red is Blue

The red line and blue line perspectives are similar in one crucial respect: both depend on the perceived beliefs and expectations of certain constituencies or groups in deciding how to allocate resources. What ultimately distinguishes them is *which* groups influence the decision. The Blue Line Imperative dictates that those whose expectations and beliefs matter most are customers, followed by employees, and perhaps, suppliers and other collaborators. In a red line culture, the line starts at the other end, with the greatest weight given to investors, typically shareholders first, then followed by debtholders.

We know that the blue line exerts a magnetic pull on the red, but is the reverse also true? It is not – at least not in general. An artificially high share price, one where the red line is well above the blue, cannot magically lift the blue line because the creation of false value is not going to "fool" the blue line into climbing as well. The only thing that can cause the blue line to rise is a true increase in value.

In fact, as noted earlier, attempts to spike "value" as defined by the red line may, more often than not, cause the blue line to fall. Not only will management expend resources that could otherwise be put toward positive-NPV activities, but when investors inevitably discover that the expected cash flows are not sufficient to support the high share price, it will drop. Causality, in any and every case, is from blue to red, not the other way around. But are there circumstances in which a healthy concern for the red line can be value-enhancing? Can red ever be blue?

There are indeed instances in which the ongoing existence of a company depends on its ability to attract cash from investors, and fast. In such cases, the expectations of potential investors become paramount, and there is no choice but to manage toward the red line, at least in the short term. Typical start-ups and companies in financial distress know this all too well. In such cases, the blue line cannot become a priority prior to assuring, at least in the short term, the survival of the firm.

To illustrate, imagine an entrepreneur with a positive-NPV idea but who lacks sustained financing. The fact that the idea is value-creating is unfortunately less relevant at this stage than the entrepreneur's ability to convince potential investors of the fact. The entrepreneur cannot afford to allot his resources to the positive-value project unless his potential investors feel the same. If he cannot make the case, the blue line is a non-starter independent of whether the idea is truly value-creating. The entrepreneur must focus on the red line because it is his only option for continued existence. For the blue line to come into play, the red line must first be taken care of.

Similarly, the survival of a company in dire financial straits may hinge on the willingness of investors to support the company through the crisis, perhaps by the exchange of debt for equity. If their expectations about the company's future cash-generating capability are underwhelming, and they aren't comfortable providing cash in the near term or swapping debt for equity, the company is forced into liquidation. In this case, as in the situation above, investors' expectations are more important than the truth represented by the underlying value, at least until the crisis is averted and the company saved. Management of the red line is again critical to having any chance of realizing the blue line in the future.

One other instance in which the direct management of indicators might be value-enhancing, involves the avoidance of a technical default on loans. This occurs when a covenant, or condition of a debt contract, has been breached and something must be done about it. For instance, a company may agree in writing to maintain a certain amount of working capital, but if they breach this condition, they are said to be in default even if all of the interest and principal payments are made on time.

This case differs from the first two in that it has nothing to do with managing investor expectations, but it is comparable because it requires the use of behaviors that would normally descend to the blue line in order to raise, or at least stabilize, it. When companies go into technical default, costly re-negotiations with lenders are often triggered. Not only must the borrower call upon expensive resources, such as lawyers and auditors, but the loan might be called in or the interest rate increased. In this way, avoiding debt covenant violations is a powerful, and potentially value-enhancing, incentive to manage earnings. Numerous academic studies have in fact shown that companies customarily take advantage of the latitude given to them under accounting rules to avoid such violations.[3]

There may be other times where colleagues or other parties suggest that management on a red line basis is appropriate, but be careful. We are reminded of a situation involving one of the oil supermajors who at that time was in the midst of divesting several businesses for the purpose of raising cash. The group CEO publicly declared that a certain dollar amount in assets had to be sold by the end of that year. The company hired external advisors to manage the sales process for one of its business units and their efforts yielded several bids from prospective buyers, all in the $400–550 million range.

The bids fell short of the value estimated by the company's own valuation specialists and a team of outside experts, who had gauged the value at around $650 million. It was agreed that the forecasts behind the valuations were reasonable, that is, not inappropriately optimistic in light of the unit's profitability and its competitive environment.

When these outside experts, none of whom had any vested interest in keeping or selling the particular business, explained their findings to the business unit manager, they were told that the business had no choice. The group CEO was committed to the policy, and so, therefore, was he.

---

[3]Examples include I. Dichev and D. Skinner, "Large Sample Evidence on Debt Covenant Hypothesis", *Journal of Accounting Research*, 2002, 40 (4), pp. 1091–1123; and J. Zhang, "The Contracting Benefit of Accounting Conservatism to Lenders and Borrowers", *Journal of Accounting and Economics*, 2008, 45 (1), pp. 27–54.

The advisors, prominent investment bankers, were highly motivated to ensure that the sale went through so that they could collect their very healthy fee. They managed to convince group executives that the "value" of a business is only that which a buyer is willing to pay, in this case not more than $550 million. The business was eventually sold for approximately that price. The internal experts and their own advisors recognized the action – selling a business with $650 million worth of expected future cash flows for considerably less – as value-destroying, but the company did it anyway. Why was the decision taken? To deliver on the CEO's objective. The company was neither in financial distress, nor dependent on investors' cash to survive, nor in danger of violating debt covenants. No value-based rationale existed for the actions of the business unit head or group executives. Be aware when the red line is a necessary evil, and be aware when it should be ignored.

Participants in our executive programs often respond to classroom discussions on the blue line by saying, *"It sounds good in theory, but it just won't work in the real world."* When they say *"the real world,"* they are referring to the everyday pressure to hit the numbers or meet the targets, and the associated fear of failing to do so.

But ask yourself this: what is so "real-world" about value destruction? In what way is it practical? What are the long-term prospects for your business if your energies are focused on indicators and other red line phenomena? Remember, value for us humans equals the stuff that makes us happy. Value for businesses equals delivering that happiness while using less energy to do it. There's nothing more practical or real-world than that.

That's all fine and dandy you say, but how do I avoid the common red line mistakes? Even if I believe in the blue line myself, how do I convince others of it? Is there a strategy for putting all this into practice?

There is.

CHAPTER SIX

# SHIFTING TO BLUE

Once a red line culture is embedded, it is indeed resilient, but hardly permanent. To break away from red line thinking and move the organization toward blue, two actions are necessary. The first is to recognize the red line behaviors in the first place. The second is to change the organizational culture to start "blueshifting" – managing toward the blue instead. You've probably heard the old bromide, *What gets measured gets managed.*" Perhaps you've said it a few times yourself. Can this expression apply in the context of a blue line that you cannot see but which must drive everything you do? We've said that the blue line represents value and that we therefore define performance by its effect on the blue line. If we can't see the blue line, how do we know what we're supposed to measure as a way of knowing whether it's moving up or down? Aren't we forever consigned to the red line no matter what we do?

No. Just as astronomers identify a wide range of cosmic objects not by seeing them directly but by observing their effect on other objects – or other objects' effect on them – the blue line can be managed without the necessity of being directly observed. Let's remember that we mustn't be swayed by those metrics that are so temptingly concrete and tangible. Part of the reason they are so tempting is that they are still vitally important, even though we are telling you that they should never be management's focal point. Share price, for example, is of course significant. In fact, we feel confident in proclaiming that those companies with the best share price performance over extended periods have been the most successful at creating value.

This still doesn't mean, however, that its observable nature makes it the correct thing to focus on. Executives manage share price in the hope that they can make it go up, but the tale of successful companies reads differently. Their share price rises because they manage toward value creation and therefore entrench behaviors throughout the organization that drive the blue line up, and along with it other measures, share price included. In directly attempting to manage share price, red line managers think they're creating value, but really they're just creating a different share price – assuming their efforts even influence share price, which often isn't the case. As we said in the previous chapter, managing the indicator is not the same as paying attention to the underlying drivers that explain *why* the indicator is saying what it's saying. It bears repeating that blue line companies do not try to "manage" indicators such as share price. They manage for value and then allow the capital market to determine how the concrete measures will shake out. Usually, they will look pretty good.

Organizations that give credence to indicators are not automatically red line, just as those that say they believe in "value" aren't necessarily blue line. The distinguishing factor is what they really manage toward. Whether we're talking about share price or any other metric, managing to achieve a specific outcome for any indicator has the same effect of diverting resources away from managing the forces that truly drive value. As we said earlier, blue line companies don't ignore indicators. What they do is use them properly.

Earlier, we started the discussion of value by taking a close look at what value is not. In the same spirit, in order to discuss what characterizes blue line management, it is important first to understand the behaviors that run counter to it.

Shortly after taking over as CEO of Daimler-Benz in 1995, Jürgen Schrempp implemented a system of performance management and incentives organized around the concept of Economic Profit (EP), also known as EVA®.[1] EP, which equals operating profit net of capital costs (invested capital × the OCC), can be a highly useful measure of performance. In fact, it can be proven that the value of any firm equals its invested capital plus the present value of future EPs. It stands to reason, therefore, that the sum of invested capital, and the present value of future EPs is math-

---

[1]EVA is a registered trademark of Stern Stewart & Co.

ematically equivalent to the expected future free cash flows discounted at the Opportunity Cost of Capital (OCC), our working definition of value. Schrempp felt that setting EP targets would surely spur value creation throughout the company. However, a funny thing happened: a few years later it was in tatters and, soon after, he was forced out as CEO.

To see where it all went wrong, we need look no further than the company's Mercedes car division. After EP targets were set in line with Schrempp's directives, division management tried to meet them via aggressive cost-cutting. As an example, an executive working for one of Mercedes' suppliers of drive trains told us that they were ordered to cut prices by 7% for a particular car model.

What do you suppose would be your reaction if your largest customer suddenly informed you that they would now only pay you $93 for every $100 they had paid you to date? The supplier, given little choice in the matter, grudgingly agreed, but that conciliation carried with it another behavior as well: a prompt reduction in the quality and thickness of the steel used in the manufacturing of the drive trains. The car division's cost-cutting goal was reached all right, but at a heavy cost. The new drive trains, instead of operating well under heavy conditions for five years as they had done in the past, now only lasted three years, and customers began to notice. In fact, they did more than notice; they started to buy other makes. A short spell of higher profits was followed by a much longer one of losses.

Where did Schrempp go wrong? He had the correct view that creating more future EP is better, all other things being equal. But this view does not imply that pumping up EP in the short term ultimately generates value for the company. Schrempp's managers did exactly as they were told. They undertook a single-minded focus on the 12-month EP targets they'd been ordered to meet, but in delivering on these targets, they were destroying value.

The story gets worse. The senior operating managers at Daimler understood that bolstering their EP targets meant destroying value. But they also knew that their bonuses, and really their jobs, depended on listening to the new head honcho. He said hit the targets, so hit the targets they did. Earlier we described the layered effect that managing the indicators has on a company, not least of which is a blow to morale once employees figure out that everyone is target-hitting in the interests of self-preservation, rather than working toward common value-creating goals. In the wake of Schrempp's directive, Daimler-Benz exhibited a classic red line climate. Morale plummeted, descending gradually to an atmosphere of loathing and mistrust. We have talked with several senior managers from the company over the years, including a few who lived through the turmoil of its disastrous takeover of Chrysler. As one person told us in describing the work environment during this period, *"It was such an ugly place. Nobody trusted anyone else. You always had the sense that you were about to get stabbed in the back."*

The most important thing to recognize is that the rot began to set in the moment Schrempp launched his "value-based" program. Nowhere was the mistrust engendered by this program more evident than in the company's handling of the delicate negotiations with Chrysler. Before the acquisition was announced, no more than a handful of people in Daimler-Benz were fully informed, in contrast to Chrysler, where hundreds of managers were actively involved. The atmosphere at Daimler-Benz was so poisonous that top management felt its people simply couldn't be trusted with strategically important information. A consequence of having only a select few people on the job, is that the job – in this case, the due diligence for the acquisition – probably isn't going to be done as well as it should be.

Our contacts inside the company tell us that while the worst excesses of the Schrempp era are over, much of the company continues to be managed on a red line basis, with senior executives tasked with hitting numbers even if it means value destruction. One senior operating manager recently told us, *"The red line express has so much momentum, I just don't see how we can stop it."*

The Daimler-Benz story is just one example among many red line misadventures. As we said, the first step in being able to manage toward the blue line is the ability to recognize the symptoms of red line management so it can be stopped in its tracks. We have set out in the following points, five of the most common manifestations of red line management played out at the top of the organization. You may recognize some of these behaviors at play in your own company. That's half the battle won.

### 1. Paying taxes on inflated earnings.

Imagine that you and your fellow managers believe the company needs to show higher revenue and earnings growth to justify its current share price. With a red line mentality, it is easy to talk yourself into assuming that analysts have embedded certain Return on Invested

Capital (ROIC), profit, and revenue-growth expectations into your share price, and that if you fall short of these expectations, the share price will plunge. So what if a shortfall is looming? Why not manufacture revenue growth and higher earnings via creative accounting practices?

You might, for example, employ "channel stuffing," pushing product downstream onto customers who aren't asking for it. Or perhaps you create fictitious revenues of the sort associated with major accounting scandals such as Enron (America), Olympus (Japan), Parmalat (Italy), and Satyam (India). To sustain the impression of higher revenues and profits, you make sure the inflated numbers are reported to the tax authorities. The company pays taxes on profits that don't even exist in order to sell the story properly.

## 2. Playing with hidden reserves.

In Chapter 4 we mentioned the practice of creating hidden reserves of accounting profits through the over-provisioning of expenses and losses. The reserves can then be released in later years to artificially boost earnings. Jack Welch, a man widely considered a paragon of value creation, was a master at this game. Soon after his retirement, he wrote a best-selling autobiography, *Jack: Straight from the Gut*. One largely ignored passage in the book reveals a disturbing attitude Welch held toward financial reporting. In the mid-1980s, GE acquired the stockbroker, Kidder, Peabody & Co. The investment turned out badly and GE was forced to take a large non-cash write-off. With just two days to go before the quarterly earnings release, here's how Welch described his management team's reaction:

> "The response of our business leaders to the crisis was typical of the GE culture. Even though the books had closed on the quarter, many immediately offered to pitch in to cover the Kidder gap. Some said they could find an extra $10 million, $20 million, and even $30 million from their businesses to offset the surprise. Though it was too late, their willingness to help was a dramatic contrast to the excuses I had been hearing from the Kidder people."[2]

---

[2]J. Welch and J. Byrne, *Jack: Straight from the Gut*, (New York: Warner Business Plus), p. 225.

To Welch, this attitude represented a prototypical team-based culture. You can practically see the tears running down his cheeks as he relates the story, contrasting the "can-do" spirit of his guys with the chronic whining of the Kidder, Peabody people.

But if you were someone who had recently been promoted to run a division at GE, what lesson would you really have learned from this? Perhaps that you had better create hidden profit reserves, so that if called upon, you could contribute in the same way your colleagues did in response to the Kidder, Peabody fiasco? Welch's primary motive was to smooth earnings in the false belief that this action delivered value. It's a move that epitomizes red line management, insofar as Welch is trumpeting the willingness of his managers to divert their efforts and resources toward the manipulation of an indicator. Welch eventually acknowledged that this kind of behavior was inappropriate and counterproductive, but not until years after he no longer held his post at the reins of GE.

3. **Citing share price impact as a means of justifying business decisions.** When someone believes he has the magical ability to anticipate how the stock market will react to management decisions, it is natural that he might also see the impact of share price as the logical argument for explaining why a certain decision is made.

This logic is not logical at all and for two reasons in particular. First, you may be a very bright person with lots of business credentials, but we're afraid you can't predict the pricing effects of your decisions any better than the next person can. Even a giant of the capital market such as Warren Buffett admits that such a skill is beyond him.

Second, remember the principal objective of your business is *not* to increase share price, but to create value. If management is taking the blue line seriously, it is focused entirely on making decisions which are right for the long-term survival and success of the company, and leaving the determination of share price to the capital markets.

An unfortunate example comes to us from the recently bankrupted Canadian telecom firm, Nortel Networks. In the late 1990s, as Nortel's share price rocketed, the company found itself in the position of trying to support the unsupportable. To keep the bubble from bursting and the share price from diving, senior management communicated guidance to the capital markets on its key indicators. Their hope was that the indicator targets would convince the analyst community that the inordinately high share price was justified. The catastrophic thinking behind this practice became evident when that guidance became the principal driver of internal target setting, even to the point of internal e-mails to employees stating the critical importance of delivering on the targets to support the share price. The markets were not impressed, and the company did not live to tell the tale.

**4. Using outcomes in the past to justify decisions in the present.**
The great economist Friedrich Hayek once said, *"Never will man penetrate deeper into error than when he is continuing on a road that has led him to great success."*[3] A more prosaic way to put this is that the past is never a guarantor of the future. Yet it is common for managers to allow historical profitability to influence capital investment. Higher capital budgets are allocated to operating units that have earned the most profits in the past, with little, or often no, thought given to the sustainability of those profits.

One of the core principles of sound financial analysis is that sunk costs – cash inflows or outflows that have already happened – are *never* relevant to proper decision-making. From a capital budgeting perspective, you should pay absolutely no heed to what a business unit's profit may have been in the past. All that really matters is what you expect it to be in the future. It is NPV and NPV only that has relevance.

A well-known insurance company's transition from strong profits to large losses over two consecutive years resulted directly from failure to heed this lesson.[4] Recent robust profits in the company's auto insurance unit led senior management to assume that if they did more of what they had done, profits would obviously soar. Additional capital resources were allocated to the unit, and strategies were devised to ramp-up the business.

One of the ideas was to cut premiums, which would therefore attract more price-conscious customers. As a result of this decision,

---

[3] F. von Hayek, *The Counter-Revolution of Science*, (Glencoe, Ill: The Free Press, 1952), p. 105.
[4] We are obliged to keep the identity of the company secret.

revenues indeed grew, though not as fast as management had hoped since the price cuts meant that each policy brought in less cash.

Even worse, the firm soon noticed an alarming spike in claims. The price-cutting strategy had drawn in high-risk policy holders, a problem known in the insurance industry as "adverse selection." In effect, the company started insuring the wrong people. As costs started to spiral out of control thanks to the number of claims, the company did what many other companies do in similar circumstances: look for ways to slash costs. Large numbers of staff were let go, including several responsible for screening new policy holders. The firewall protecting the company from bad risks had been breached, and there was no one to fill the gap. Inevitably, things went from bad to worse.

### 5. Practicing sale-and-leaseback.

The process of sale-and-leaseback is quite simple: the owner of an asset sells it, then leases it back from the buyer. Like any capital transaction, such a move can translate to positive or negative NPV. This in itself doesn't put it in the category of red line behavior. However, experience shows us that the most common motives behind such deals have little or nothing to do with value creation.

Browse the websites of the many finance companies that structure sale-and-leaseback transactions and you notice almost interchangeable selling propositions. First, the leaseback is proposed as a way to free up the owner's capital while allowing her to retain possession and use of the property. This sounds fine in principle, but dig deeper and you'll realize that the move is just an attempt to generate more debt with the underlying asset serving as collateral for the loan. Leases are just like mortgages – secured (i.e., collateralized) forms of borrowing – and therefore have the same effect on debt capacity, dollar for dollar, as do other types of borrowing. Leasebacks may allow a company to remove debt from its balance sheet, but as long as bankers and other lenders aren't completely daft, leasing does not increase debt capacity. To use leases to get debt off the balance sheet is pure red line management.

Furthermore, leasebacks are advertised as ways to dress up other key accounting numbers. One effect of removing debt (and the related asset) from the balance sheet is to increase return on capital measures, such as ROIC. Some CFOs are attracted to leases because of the belief that certain ROIC expectations are built into their company's share price. Higher ROIC, the thinking goes, should translate into higher prices, even if the increase in ROIC is the result of an accounting policy and not higher cash flows.

Although you won't see it advertised, there may be perceived benefits from leasebacks on the Statement of Cash Flows (SCF) as well. To see why, consider what happens when a company takes out a direct loan from a bank. The resulting inflow of cash is classified as a "financing" cash flow. However, if the loan is structured as a sale and leaseback, the company might elect to classify the cash received from the asset sale as an "investing" cash flow. The perceived advantage to the company is that analysts might increase their estimates of free cash flow. Free cash flow equals the cash generated from operating activities net of capital investment, with investment reduced by the proceeds from the sale of property, plant, and equipment. In other words, capital investment equals the amount invested in new assets, *net* of the cash received from the sale of old assets. Therefore, classifying cash receipts from sale-and-leaseback deals as investing cash flows will reduce net capital expenditure on the SCF, and increase the estimate of free cash flows.

Any analyst worth his salt won't be fooled by this ruse, of course. Regardless of how the company classifies the cash received from the asset sale and leaseback, it should never be included in estimates of free cash flow. It is a financing cash flow, pure and simple. But that's not really the point. If company executives believe that those in the investing community believe their ability to generate cash from the business is influenced by the way they classify cash flows from financing events, they're content – no further motivation is necessary.

A few years ago, researchers at the Financial Analysis Lab at the Georgia Institute of Technology discovered several instances of American companies having taken advantage of the laxity under accounting rules and classifying the cash received from leasebacks as investing cash flows.[5] In some cases, including the national pharmacy chain CVS Corporation and movie theatre owner AMC Entertainment, a simple

---

[5] http://www.cfo.com/article.cfm/3499792?f=related

re-classification of the cash as financing cash flows would have caused free cash flow to go from positive to negative. The Lab's director was quick to point out that he found no evidence of ulterior motives behind these actions, but we suspect that the desire to dress up reported cash flows may have played a role in the financing decision of at least some companies.[6]

We are not opposed to sale-and-leasebacks transactions *per se*, but as with any capital allocation decision, only one question matters: is the decision positive-NPV? If you cannot confidently answer this question in the affirmative, with cold hard facts to back up the answer, your company will almost certainly be better off looking for other ways to finance the operation.

## The Indicator Illusion

The previous section outlines the sort of red line behaviors commonly found at corporate level and with top management in general. But make no mistake – red line management is a virus that can spread through all levels of a business. As we explained in Chapter 5, the main culprit is the misuse of performance indicators. To initiate the shift to blue in your organization, it is vital that you recognize red line behavior wherever it may be happening, and in addition that you can explain to the person doing it why it is not, in fact, creating long-term, sustainable value for the company.

To do this, you must be able to discuss indicators and clarify their proper place as tools of learning rather than focal points in and of themselves. It may take more than one try to make the explanation stick. In that same spirit, allow us to spend a little more time on indicators as well, given their overall importance to the Blue Line Imperative.

---

[6]As of this writing, the US Financial Accounting Standards Board and the International Accounting Standards Board are debating the issuance of a joint standard that would eliminate, or severely curtail, the use of operating leases. Nearly all leases will henceforth have to be capitalized, which means that the present value of future lease payments will appear on the balance sheet as both an asset and a liability. It will be interesting to see how the leasing industry is affected by this change.

As we said, the first problem with indicators is that they sometimes tell you the wrong thing, or they don't tell you what you think they are telling you. You might think of the different indicators in your car. The speedometer indicates the speed you're traveling, while the fuel gauge tells you how much fuel you have left. In other words, don't look at the fuel gauge if you want to know whether you're speeding. Be wary of what you believe – if the fuel gauge gets stuck, you could face a long walk with a gas can in your hand. If you don't know what the indicator is telling you or how reliable it is, you could end up a long way from home.

In the popular film *Die Hard 2*, the bad guys take over an airport and threaten to crash all the planes coming in to land. They demonstrate their ability to do this by remotely altering the ground level setting on a plane's radar altimeter. The unlucky movie pilots believe they are still some way above the ground only to crash into it in spectacular special-effects fireballs. Of course, this is an action film and flames look much better in the dark, so the fictional pilots don't have the option of looking out of the cockpit window and seeing where the ground really is. If you believe your indicator tells you something and you dogmatically stick to that belief without checking out of the window every now and then to see if anything out there is telling you something different, you might be in for a bad ending just like those poor suckers in the cockpit.

We don't have to look to the silver screen to see the dangers that can arise from the management of indicators; we can look to Argentina instead. In 2011, the Argentine Government filed charges against the managers of nine economic consulting firms for circulating inflation estimates that challenged the figures produced by Indec, the national statistics agency.[7] The charges – for producing "false figures" and "generating uncertainty among consumers" – carried heavy fines and possible jail sentences for the firms' directors. Inflation rates are a touchy subject for Argentina's rulers for several reasons. First, high inflation implies heavy political costs, largely due to toxic memories from Argentina's hyperinflation in the 1980s. Second, lower inflation rates mean higher reported real

---

[7]Taos Turner, "Argentina Charges Economists", *The Wall Street Journal*, July 9 2011.

GDP growth, which makes the government's economic policies look more effective. And third, the government can save billions on payments to holders of inflation-linked bonds. The ruling party was particularly sensitive to these concerns throughout 2011 because of planned elections for October of that year.[8] The discrepancies between the official inflation estimates and the ones produced by private consultants can be traced to the decision in 2007 by then-President Néstor Kirchner to replace long-serving civil servants at Indec with loyalists of the ruling party.

The story turned from tragedy (for the unfortunate consultants) to farce when the government targeted, as the embodiment of its view, the price of the Big Mac, the iconic offering of McDonald's. In 1986 *The Economist* began publishing its now famous Big Mac index as a way to measure inflation and, drawing on the concept of purchasing power parity, the extent to which a country's currency is over- or undervalued. Designed partly in jest, the longevity of the index shows that it has come to be seen as a useful barometer of price stability. For this reason, the government was embarrassed by an analysis in the magazine that compared the prices of a Big Mac between 2000 and 2010 with official inflation rates covering the same period.[9] According to the index, prices rose in Argentina by an average of 19% per year, versus average annual price increases of just 10% according to official estimates – the largest such discrepancy in the world.

The country's economy minister then "asked" McDonald's to lower the price of the Big Mac, and the company promptly complied. The Big Mac is now noticeably cheaper than competing products, such as Burger King's Whopper, which isn't the subject of government scrutiny. You can get a Big Mac if you really want one, but because the price is artificially low, it's no longer promoted. Instead, McDonald's tries to steer its customers to other products that aren't so politically sensitive and which, therefore, may be priced at whatever the market will bear. Despite all this, the government continues to insist that inflation is not a problem. Or if it is a problem, the fault lies with the businesses that raise prices – not with government policy, of course.

---

[8]Which the ruling party duly won.
[9]http://www.argentinaindependent.com/tag/imf

## Goodhart's Law

Charles Goodhart, former Chief Advisor to the Bank of England, described a related problem with indicators in a way that has since come to be known as Goodhart's Law:

> "Any observed statistical regularity will tend to collapse once pressure is placed upon it for control purposes."[10]

In formulating his law, Goodhart's primary concern was the use and misuse of indicators in controlling inflation and other aspects of monetary policy. However, his law applies to performance measurement systems generally. In just about any setting, when a measure becomes a target, it ceases to be a good measure. Essentially, Goodhart's law is an economic, political, and sociological analogue to Heisenberg's uncertainty principle in quantum mechanics, which tells us that by measuring a system, any system – whether we're bouncing light off an electron or determining operating profit – we usually disturb it.

To illustrate this problem, researchers A. Rosga and M. L. Satterthwaite wrote about the use of indicators in measuring compliance with human rights standards agreed by international bodies such as the United Nations. At a gathering of officials and consultants on the subject of human trafficking, one participant described her efforts to gather field data in this way:

> "My first effort at collecting information was easiest because no one knew what I was talking about. My second effort was harder, because they [government agency representatives] were anticipating me. The third was worse because they were deliberately creating reality to suit their own ends."[11]

---

[10] http://www.atm.damtp.cam.ac.uk/mcintyre/papers/LHCE/goodhart.html. Goodhart formulated his "law" in the 1970s.

[11] AnnJanette Rosga and Margaret L. Satterthwaite, "The Trust in Indicators: Measuring Human Rights", Berkeley Journal of International Law, 2009, 27(2), p. 287.

In other words, once local government officials understood her mission and thus realized what the indicators were for, they "painted a picture" that they believed would enhance their country's standing in the international community with no thought as to whether that picture had even a passing resemblance to the truth.

Rosga and Satterthwaite offer a second example, this time focusing on efforts to improve educational opportunities for girls.[12] A common indicator in this area is the ratio of girls to boys enrolled in primary education. Because countries are rewarded for demonstrating ratios close to one in this regard, there is a strong incentive to document female school enrolment.

The problem is that such figures say nothing about whether the substantive right to an education has been fulfilled. Female enrolment may be high, but are girls actually attending school? And if they do attend, how are they treated when they get there? And what is being taught? The paradox is that as policy-makers identify metric after metric offering something concrete to measure, those being measured, once they know they are, will gradually learn how to manipulate them. The indicator showing that more girls are enrolled in the education system may seem to show progress, but the real question – whether the underlying human right is being served – is no closer to being answered.

Goodhart's Law reinforces the idea that not only do humans adjust their behavior based on the indicators they're aware of, offsetting any effectiveness the indicators may have in guiding policy in the first place,[13] we also, and perhaps more dangerously, tend to rely on indicator targets as a primary means of doling out rewards.

David Hume, one of the leading figures of the eighteenth-century Scottish Enlightenment, warned of the inherent contradiction between "is" and "ought," the former being a description of the current state of the world on some dimension, the latter an aspiration of what the state

---

[12] Ibid., p. 286.

[13] Readers may notice the similarities between Goodhart's Law and the theory of "Rational Expectations" associated with Robert Lucas, Thomas Sargent, and others. We use Goodhart's formulation because it speaks more directly to the use of indicators.

of the world *should* be. When using an indicator as a tool for accountability, we are not only asking it to *describe* but also to *prescribe*.[14] We confuse what *is* with what *ought* to be, and in the process render the indicators all but useless as descriptors of reality. As Keith Hoskin writes,

> "It is not just system-beating, degenerate or unethical self-interest that keeps [Goodhart's] law in place. . . . It is the logical contradiction within which we are all placed by being subject to measures that are also targets, regardless of our degree of individual willingness to adopt system-beating practices."[15]

To put it another way, the problems with indicator-based management persist even when the intentions of the people being evaluated by the indicators are entirely honorable. These problems are universal, applying to nearly all forms of human endeavor, including government and non-profit organizations. In fact, in such arenas we often observe the indicator problem being more egregious than in the private sector, since private companies are at least subject to market constraints. Humanity tends to kill value-destroying businesses a lot faster than it kills value-destroying governments or political regimes. This instinctive leniency – we want to believe that governments and other political or civic entities are doing good, so we, as the market, give them a longer leash – explains why one doesn't have to look far to find examples of red line behavior in the non-profit sector. Here are four among many.

### What Gets Measured Gets Managed (1)

Writing in the UK *Observer* newspaper in 2008, Simon Caulkin high-lighted an ill-fated example of managing what gets measured in the UK's National Health Service. Efforts in a particular hospital to meet a specific indicator target – namely, reducing emergency room waiting times – appeared

---

[14]Keith Hoskin, "The awful idea of accountability': inscribing people into the measurement of objects", in R. Munro and J. Mouritsen (eds.), *Accountability: Power, Ethos and the Technologies of Managing*, (London: International Thomson Business Press, 1996), p. 267.

[15]Ibid., pp. 278–279.

to have been highly successful, but there was one slight problem: mortality rates for emergency heart attack admissions had gone up.

Reporting on the situation in the *Economic Journal*, Caulkin quotes the research of Professor Carol Propper and colleagues, *"It seems unlikely that hospitals deliberately set out to decrease survival rates. What is more likely is that in response to competitive pressures on costs, hospitals cut services that affected [heart-attack] mortality rates, which were unobserved, in order to increase other activities which buyers could better observe."*[16] In short, patients didn't have to wait as long before seeing a doctor, which, through a red line lens, was a highly positive result – until you counted the fact that more of them died waiting.

### What Gets Measured Gets Managed (2)

The New York Police Department has lately been at the center of a controversy over two targets: first, increasing the number of stop-and-frisk searches in order to create the perception of greater security, and second, decreasing the numbers of specific types of crimes to make the overall picture look rosier. In both cases, precinct commanders were attempting to meet targets set for them by their bosses.

To hit the first type of target, police were allegedly stopping New Yorkers on a regular basis and searching them whether or not they had reasonable cause to do so. To hit the second, precincts were supposedly making a habit of *"downgrading felonies to misdemeanors"* and *"actively discouraging crime victims from reporting crimes."*[17] Anything to meet the targets.

Similar actions have been reported in police departments in Europe and Asia, all in the interest of making key metrics look better. The Tokyo police department, for example, reports an astonishingly high success rate for solving murder cases – 96% to be precise. How do they do it? Very cleverly: by recording murders on the official ledger only if the culprits are found. Unless they arrest the perpetrator of the crime, the death in question goes into the files as an accident or abandoned body.[18]

---

[16] Cited in http://www.simoncaulkin.com/article/74/.

[17] R. Balko, "Oh, You Mean Those Quotas", *Reason* magazine online, May 17 2010. http://reason.com/archives/2010/05/17/oh-you-mean-those-quotas

[18] As reported in the movie, *Freakonomics*, based on the best-selling book of the same name.

That's certainly one way to meet the target, but it isn't exactly adding value.

## What Gets Measured Gets Managed (3)

An elementary school in Houston, Texas, was investigated after some of its state achievement test results appeared to be far better than the district would have anticipated. As one observer pointed out, *"For the principal and assistant principal, high scores could buoy their careers at a time when success is increasingly measured by such tests. For fifth-grade math and science teachers, the rewards were more tangible: a bonus of $2,850."*[19] An investigation into the matter led to the resignation of the principal, her assistant, and three teachers, and the reported scores were invalidated.

A similar investigation has tarnished the legacy of the former head of Atlanta's public school system. Ironically, she had been named US Superintendent of the Year in 2009, largely on the strength of the system's reported gains in test scores. Unfortunately for her, behind the story of the impressive increase in scores was a different tale unearthed by state investigators: that of teachers and principals in 44 different schools erasing and changing test answers.[20]

Contributing to the problem were unrealistic test-score goals. A culture of pressure and retaliation had spread throughout the district, and because each time a school attained one target, the next would be raised, so pressure was constantly on the rise as well. Cheating one year created a need for more cheating the next. *"Once cheating started, it became a house of cards that collapsed on itself,"* the investigators said.[21]

In covering still another cheating scandal, this time in the state of Pennsylvania, a *New York Times* reporter wrote, *"Never before have so*

---

[19]T. Gabriel, "Under Pressure, Teachers Tamper with Test Scores", *New York Times* online, June 10 2010. http://www.nytimes.com/2010/06/11/education/11cheat. html?_r=1&src=mv&scp=3&sq=teachers%20tamper%20with%20test%20 scores&st=cse

[20]http://www.csmonitor.com/USA/Education/2011/0705/America-s-biggest-teacher-and-principal-cheating-scandal-unfolds-in-Atlanta

[21]http://www.ajc.com/news/investigation-into-aps-cheating-1001375.html

*many had so much reason to cheat. Students' scores are now used to determine whether teachers and principals are good or bad, whether teachers should get a bonus or be fired, whether a school is a success or failure.*"[22]

## What Gets Measured Gets Managed (4)

Beginning in 1993, local governments in the UK were required by the national government to share with the populace over 150 targets that they would be mandated to hit relating to a broad range of public services. The idea seemed perfectly logical and bright: make the performance of local authorities more transparent, and give officials an incentive to improve service to their constituents.

But the idea was fatally driven by red line thinking, and the results corresponded. Marilyn Strathern, an anthropology professor at Cambridge University, cites the example of food services to the elderly. Although most older people expressed a clear preference for frozen or microwaveable meals as opposed to food delivered by Home Helps (a "meals on wheels"-type service in which local council employees deliver meals in person), "number of home helps" was the indicator target the national government gave the municipalities to hit.

Strathern writes, *"An authority could only improve its recognized performance . . . by providing the elderly with the very service they wanted less of, namely, more home helps."*[23] So that's exactly what they did. The regrettable outcome: *"The language of indicators takes over the language of service."*[24]

And why not? As we've mentioned, if your performance objective is based on an indicator, and your bonus (and maybe even your job) is linked to that indicator, what other choice do you have? You're going to put blinders on to focus on that indicator and little else. You're going to have tunnel vision, because only the target you've been given has any consequence as far as you're concerned. You're going to do everything

---

[22] http://www.nytimes.com/2011/08/01/education/01winerip.html?hpw

[23] Marilyn Strathern, "The Tyranny of Transparency", *British Education Research Journal*, 2000, p. 314.

[24] Ibid.

you can to nail it, even if it means value destruction and the total absence of worthwhile learning. As Upton Sinclair wrote nearly 80 years ago, *"It is difficult to get a man to understand something when his job depends on not understanding it."*[25]

Let's say it a different way. If you're managing the indicator and not what it's supposed to indicate, you're not managing for value creation. And recall, once you start managing the indicator, it is no longer reliably indicating what you thought it was telling you in the first place, creating a vicious cycle. The bad guys in *Die Hard 2* managed the radar indicator, and it certainly wasn't telling the pilots where the ground was.

One of us, Kevin, had the idea as a child to manipulate his temperature when he didn't feel like going to school, by sticking his forehead close to a lamp. His mom felt the hot forehead and took it as an indicator that little Kevin clearly had a fever. Never mind that the fever would miraculously disappear soon after school was finished for the day. Mom thought the indicator was measuring something it wasn't, and Kevin was playing her false interpretation for all it was worth.

The easier way is often the one chosen, but it's seldom the one that leads to better results. Any indicator target can be obtained by either creating value or destroying it, and the latter is inevitably going to prove easier. Let's assume a manager has been given the task of reaching 15% market share or else he won't get his bonus. Granted this sounds extreme, but imagine that the market share target can be achieved either by spending $100 million on marketing, PR, advertising, and working hard, or by spending $1 billion and not working so hard. What will a manager who tends toward red line thinking do? He will accept the billion-dollar budget, spend it happily, then sit back and wait for his bonus while lots of company value goes down the tube. A manager who upholds the Blue Line Imperative will spend the $100 million judiciously, do his job the best he can, probably earn the bonus anyway, and in the process generate value for the organization.

---

[25] Quoted in Michael Shermer, 2011, op cit., p. 266. The quote can also be found in Upton Sinclair, *I, Candidate for Governor, and How I Got Licked*, (New York: Farrar & Rinehart, 1935).

Be wary of this contrast and try to identify it in yourself or others around you. It will always be easier to hit indicator targets via the value destruction route instead of the value creation one. Managers who focus on performance objectives based on indicators often meet their targets but destroy value as a result. Worse, for reasons discussed, their behavior disheartens other employees. Since reaching their target means their own promotion and bonus, and because it will be obvious to others that the value-destroying manager overspent in order to ensure those outcomes – thereby reducing overall profits and ruining the employees' chances of a raise on the strength of a good year – the behavior, and the result, will justifiably be seen as unscrupulous. We are sorry to say that in too many cases the manager doesn't care. That's not quite accurate, actually. It isn't that he doesn't care. It's that he would fail to see why anyone else would be upset at the result. After all, he met his targets, increased market share as he was mandated to do, and earned his bonus fair and square. In a red line culture, it is very difficult to pin it on one person. The Blue Line Imperative demands a shift in thinking up, down, and across the organization. Where the blue line is understood and embraced by everyone in the company, and red line mentality is eradicated for good, value creation begins.

# CHAPTER SEVEN

# THE HAZARD OF GROWTH

The number "29" seems harmless enough, but it got General Motors into a heap of trouble not too long ago. In 2003, GM's North American market share dropped to 28.3%. This was down from 28.7% the previous year, the first such decline the company had experienced in three years. So what did the GM executives in their wisdom choose to do? Did they take a closer look at the value drivers within the company in order to figure out where greater value generation could be achieved? Did they adopt a fresh perspective regarding the manner in which they judged the Net Present Value (NPV) of the projects they invested in or the day-to-day decisions they made?

No – they started wearing lapel pins in the shape of the number 29. The purpose of this move was, of course, to challenge and encourage employees to push GM's market share back over that figure. When an indicator falls, what does the red line view tell you to do? Set it higher as a target for next time. The president of GM North America was quoted as saying, *"29 will be there until we hit 29. And then I'll probably buy a 30."*[1] We don't need to tell you how that turned out, but we will anyway: value-destroying growth investment, a slow death resulting in bankruptcy, and a massive bailout from the US government.

A growth strategy can be an inspirational thing, but it can also be dangerous. Nowhere does management of the red line manifest itself in a more noxious way than in obsession with growth and market share.

---

[1]H. Simon, F.F. Bilstein and F. Luby, *Manage for Profit, Not for Market Share*, (Boston: Harvard Business School Press, 2006), p. 1.

We suspect that if you could somehow observe the growth targets set by senior managers in every publicly traded company in Europe, North America, and Japan, and weighted the targets for firm size, the result would be a lot greater than the 2% annual growth one might reasonably expect for the global economy. The conclusion is that many corporate growth plans are doomed to destroy value. Why?

To see better how an undue focus on "growth" can cause companies to derail, consider Figure 7.1 below which shows that the availability of investment projects with expected returns (Return on New Invested Capital or RONIC) that exceed the Opportunity Cost of Capital (OCC), tends to diminish over time as the effects of competition, along with the demands of customers, employees and suppliers, cause profits to reliably erode. If you are a company focused on gaining market share, you must woo customers from your competitors. To make potential customers aware of your existence, you will need to expend resources. To make it worthwhile for those customers to switch – by improving the product say, or lowering the price – you will need to expend still more. As long as the expected return from pursuing these customers exceeds the OCC, your investors will be happy to provide the needed funds.

Then again, industry rivals might conspire to limit competitive forces before the RONIC is driven all the way down to the OCC. As an example, if we assume the OCC is 8% and the average RONIC 12%, it wouldn't surprise you if your competitors were to form a cartel to stave off the threat of you entering the fray.

Will it work? For a while maybe, but eventually nature will exert itself and bring the cartel down. Customers today are more sophisticated than ever before, and it won't take long before they recognize that firms in the industry are earning RONICs above the OCC and then demand that they start using these extra profits to improve product quality, reliability, service, price, or some combination. Should the cartel fail to respond, we all know what the result will be. The market, in its instinctive wisdom, will pave the way for new entrants with value-creating ideas. The incumbents will not be able to ignore this new attack; they will respond to the pressure by driving the RONIC to 8% in spite of their initial efforts, along with those of their fellow colluders, to keep it higher.

It isn't only consumers who will take a stand against the scheme. Employees too, witness the attractive returns delivered to investors and

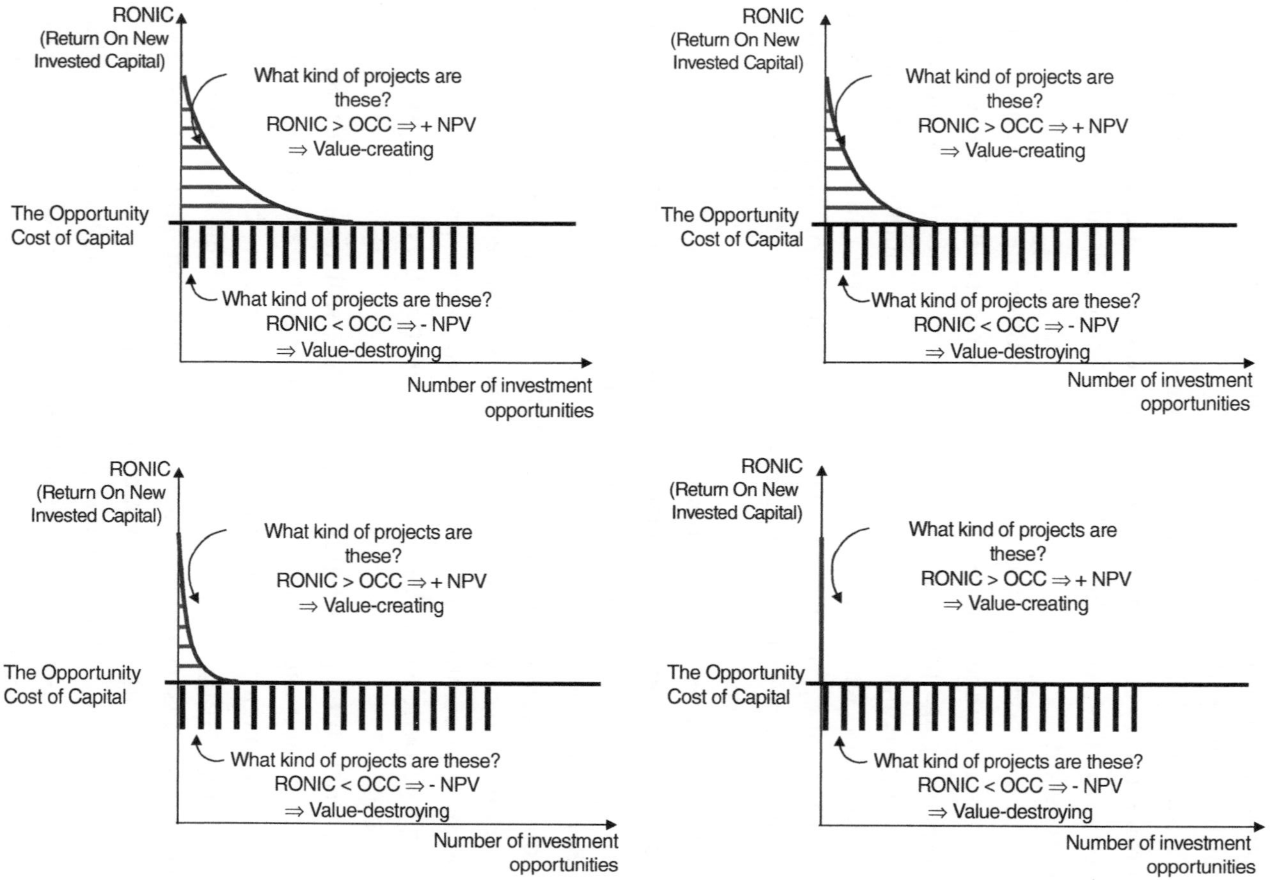

Figure 7.1: Competitive/economic forces squeeze out value-creating opportunities

speak up, wanting their fair share of the pie. They demand pay increases, further driving down the RONIC.

And what about suppliers? Will they not want to increase their own profits if they know how much money the company is generating? Why, certainly they will. The RONIC takes another hit and is pushed further downward.

In understanding and embracing the Blue Line Imperative, you must always remember the lesson from Chapter 1: that value creation is a conspiracy against nature – not just the birds, trees, rivers, and lakes, but all living organisms, including your customers, employees, suppliers, and competitors. Whenever you manage to achieve a rate of return greater than the OCC, you may as well paint a target on your chest, because nature does not share your idea. It will always respond by trying to take profits from you.

This is why superior returns can never last, and why the narrow pursuit of "growth" is foolhardy. The market does not care about your, or anyone else's, desire to buck the natural order. Nor, in the end, do your customers, employees, suppliers, or competitors. If they can benefit from forcing your company's RONIC below the OCC, they will do it without thinking twice. Your customers will have few qualms about pushing you into bankruptcy if it means more profits for them. They might experience a few pangs of guilt, but they'll get over it quickly enough.

And nature, in its reflexive knowledge, will again adjust. Since investors put their capital into a business to earn at least the OCC, then if the forces wrought by competition drive this return below the OCC, they stop supporting the company. The normal result is a combination of consolidations (mergers and acquisitions) and liquidations, until the forces have balanced out again and the RONIC is driven back up to, or above, the OCC. This explains the frequent phenomenon of "roll-ups" in declining industries. An industry that cannot offer competitive returns to 12 companies, for instance, may be able to do so if it consolidates into three larger firms. It will do whatever it must to attract future investment, and to do this it must find a way to offer the prospect of returns at least as great as the OCC. The continuous interplay of the various forces of nature mean that the steady state outcome – one that balances the demands of customers, employees, suppliers, and investors – is one where the RONIC equals the OCC.

Recall our earlier example of fiddling with the shower knob to try to achieve just the right temperature. Nature is forever making a similar adjustment, so whether the RONIC yo-yos up or down, nature will seek to pull or push it back toward the OCC, because that's the only path toward value and happiness.

## Competitive Advantage

If every element in nature is trying to maintain balance toward the OCC, how can any company consistently earn expected returns above it? We've just said that any returns your company generates above the OCC will inevitably be taken back by nature in order to restore the value creation balance. So aren't we contradicting ourselves?

The answer lies in SCA, or Sustainable Competitive Advantage. When your company establishes this it creates the possibility – not a guarantee, but a possibility – of earning an expected return exceeding the OCC. Looking again at Figure 7.1 above, we see room in the curve between the RONIC and the OCC. Companies that exist within that space in the curve have developed some trait or attribute that enables them to push vigorously back against the competing forces and keep them at bay, while still creating value. This is the Blue Line Imperative in action. A company with no discernible SCA will inevitably be overwhelmed by nature's proclivity, driving its RONIC down to the level of the OCC or below it. Creating some form of SCA will allow it to find space under the curve, and within that space, to create the possibility of positive-NPV investments.

Not many firms in an industry or given market can have an SCA at any point in time. Just as in an Olympic event, only one competitor can win gold, only one can win silver, and only one can win bronze. Everyone else is shut out despite being world-class in their field. So too, is the nature of competition, and thus competitive advantage in business. If the industry is always going to be forced by nature toward a state in which expected earnings are equal to the OCC, and if only one or two companies in a given market can have an SCA that allows investments where the RONIC is greater than the OCC, then logically for the other companies in the industry, the curve only offers the other kind of project in which expected returns on new investments fall below the OCC. Otherwise, the average RONIC for the industry average would exceed it. For most companies, in other words, the only way to achieve "growth" is to invest in negative-NPV projects.

Let's say it another way. If your company does not have a competitive advantage, it has no access to positive-NPV projects. Therefore growth can come only from value-destroying projects. Since any additional investment is going to necessarily destroy value, the obvious move is for the company to stop investing and, ideally, divest and shrink.

The managers in your company are going to be pretty reluctant to take this course of action naturally, but if they don't, nature – that is, your customers, your competitors, and your investors – will surely compel them to. But if your company has unlocked some type of SCA, that is, it has created access to ongoing positive-NPV investments, value-creating growth is possible. It still depends on a blue line mentality, however. If you're the CEO and you decide to set a growth target above that which is achievable via your positive-NPV investments, the result of your recklessness will be value destruction. But you know that by now.

SCA doesn't last forever. That's why we call it sustainable and not permanent. When Sony's Walkman portable audio cassette player came out in the 1980s, it took the industry by storm and was spectacularly profitable for its maker. However, no one could predict how long the trend would last. Certainly even the best market prognosticator couldn't have told you in 1986 that the device would eventually come to be replaced by the MP3 player. Like a shooting star, the competitive advantage was there one moment and gone the next. New ideas come along, technology progresses, and preferences change.

The rule holds not just for products, but at a company level too. History is littered with stories of firms that have rocketed to the top and then crashed to earth. Some periods of advantage last longer than others, of course. The British East India Company could apparently earn consistently high profits for over 250 years, but technology and change eventually caught up with even this behemoth, which was finally dissolved in 1873.

Today, even the most successful companies are hard-pressed to sustain competitive advantage for more than a decade or so. Nature simply won't allow it. Tom Peters and Robert Waterman famously pointed out that they *"weren't writing Forever Excellent"*[2] in their note to the 2004 edition of their seminal book, *In Search of Excellence*. Studies by Wiggins and Ruelfi[3] found that a staggering 95% of the authors' sample of 6,772 companies failed to hold their position of advantage for 10 years. Less than half a percent stayed the course for 20 years, and just three were still out-performing their rivals after 50.

The researchers discovered another, equally interesting, fact. Not only was competitive advantage difficult to sustain, it became progressively more difficult as time went on. The more recent the observations, the lower the average time companies were able to retain their competitive advantage, and the higher the chances of their being overtaken by competitors. In short, sustaining competitive advantage has become a tougher test than ever.

Rather than defining competitive advantage in quite the same way as Wiggins and Ruelfi, we choose to employ a more rigorous standard represented by a single question, which will come as no surprise: is the company creating value or not?

This question is one that should be self-directed. The lens is turned only inward, not outward, since looking at others will give you no useful information. We have observed numerous companies making the fundamental mistake of defining competitive advantage according to a comparison against other firms within the same industry. Throughout the strategy literature we come across statements such as, *"An attractive position is one where the company has a sustainable competitive advantage enabling it to earn greater profits than its industry peers."*[4] It is our hope that at this point, you are automatically saying to yourself that this

---

[2]T.J. Peters and R.H. Waterman, *In Search of Excellence*, (London: Profile Books, 2004), Authors' Note: Excellence 2003.

[3]R.R. Wiggins and T.W. Ruelfi, "Sustained Competitive Advantage: Temporal Dynamics and the Incidence and Persistence of Superior Economic Performance", *Organization Science* 2002, 13(1), pp. 81–15. R.R. Wiggins and T.W. Ruelfi, "Schumpeter's Ghost@ Is Hypercompetition Making the Best of Times Shorter?", Strategic Management Journal, 2005, 26, pp. 887–911.

[4]E. Beinhocker, *The Origin of Wealth*, (Boston: Harvard Business School Press, 2006), p. 324.

definition is purely red line, that is, deeply misleading and potentially dangerous. We are all a part of the global value chain. Value creation does not occur *because* we outperform our competitors. That this is a natural by-product of value creation does not mean it is the straw that stirs the drink. If an industry is filled with companies pursuing negative-NPV investments, that is, collectively destroying value, there is still going to be some company at the top of that list – the company doing the least bad. If no one in this imaginary industry, including the industry "leader," is capable of earning at least the OCC on new investments, the only way the industry could add any value would be to contract. Merely doing better than one's peers says nothing about whether future investments will create value or destroy it. Similar to what you might tell your kids, the Blue Line Imperative demands that you measure yourself only against yourself.

So markets become increasingly competitive over time and companies find it progressively more difficult to identify and exploit ways to create true value, that is, earn rates of return on new investments above the OCC. Competitive advantage is ephemeral and everything around us conspires to keep it that way. Is this a bad thing? Hardly. It is a victory for humanity every time a company loses its competitive advantage because every such moment signals that the forces of nature are firing on all cylinders, working to demand an ever-more-efficient use of our scarce resources and, therefore, greater happiness.

## Growth Versus Value

So should your company simply ignore growth as an objective? Should it be summarily discarded as a tool of evil and disaster? Not at all. The issue is that most firms, even those in mature sectors, express growth as a major performance objective – whether defined in terms of revenues, market share, profits, cash flow, total assets, or some other indicator – in a self-contained way. Setting growth as a target in isolation will indeed lead to problems. The question must be reframed: it should never be simply, *"How much can we grow?"* but rather, *"How much can we grow without destroying value?"* The answer depends entirely on the expected returns you can deliver on new investment, which in turn depend on the strength of your sustainable competitive advantage.

Consider the situation of a company we will call Claringdon Glassworks, a firm that has established sufficient competitive advantage over the past few years to generate value-creating revenue growth of 6% per year:

Table 7.1: Claringdon Glassworks' financial results for the past three years

|  | Past 3 Years | | |
| --- | --- | --- | --- |
|  | 2010 | 2011 | 2012 |
| Revenues | 864 | 907 | 952 |
| Cost of Goods Sold | (432) | (454) | (476) |
| Selling, General & Admin | (130) | (136) | (143) |
| Depreciation | (143) | (150) | (157) |
| EBIT | 160 | 168 | 176 |
| Taxes on EBIT | (56) | (59) | (62) |
| NOPAT | 104 | 109 | 115 |
|  |  |  |  |
| Depreciation | 143 | 150 | 157 |
| Increase in Working Capital | (5.0) | (5.5) | (6.2) |
| Capital Expenditures | (177) | (189) | (199) |
| Free Cash Flow | 64 | 64 | 66 |

By early 2013 (when this analysis takes place), Claringdon's competitive advantage is beginning to erode, and this backslide can be seen in the numbers over the explicit forecast period. The first noticeable impact is that growth in future revenues, profits, and cash flows has declined despite previous levels of Capital Expenditures (Capex) remaining unchanged. A typical response to this loss of competitive advantage is to invest even more in the hope that it can be restored. As we see below in Table 7.2, this is a mistake.

As you can observe in the forecast in Table 7.2 below, the returns delivered on investments in the past three years were 13.2%, 12.2%, and 11.9%. By 2013, it is becoming clear that Claringdon's competitive advantage is slipping, but it continues to invest. Because of its deteriorating competitive position, the additional future Earnings before Interest and Tax (EBIT) and Net Operating Profit after Tax (NOPAT) the company realizes from new investments is not as high as it was in earlier years.

Table 7.2: Claringdon Glassworks' recent financial performance and forecast

**Forecast – Income Statement/Balance Sheet/Cash Flows**

| | Past 3 Years | | | Explicit Forecast Period | | | | | |
|---|---|---|---|---|---|---|---|---|---|
| | 2010 | 2011 | 2012 | 2013 | 2014 | 2015 | 2016 | 2017 | 2018 |
| Revenues | 864 | 907 | 952 | 1,000 | 1,040 | 1,082 | 1,125 | 1,170 | 1,217 |
| Cost of Goods Sold | (432) | (454) | (476) | (500) | (520) | (541) | (562) | (585) | (608) |
| Selling, General & Admin | (130) | (136) | (143) | (150) | (156) | (162) | (169) | (175) | (182) |
| Depreciation | (143) | (150) | (157) | (165) | (172) | (178) | (186) | (193) | (201) |
| EBIT | 160 | 168 | 176 | 185 | 192 | 200 | 208 | 216 | 225 |
| Taxes on EBIT | (56) | (59) | (62) | (65) | (67) | (70) | (73) | (76) | (79) |
| NOPAT | 103.9 | 109.1 | 114.5 | 120.3 | 125.1 | 130.1 | 135.3 | 140.7 | 146.3 |
| | | | | | | | | | |
| Depreciation | 143 | 150 | 157 | 165 | 172 | 178 | 186 | 193 | 201 |
| Increase in Working Capital | (5.0) | (5.5) | (6.2) | (9.5) | (9.9) | (10.3) | (10.7) | (11.1) | (11.6) |
| Capital Expenditures | (177) | (189) | (199) | (209) | (217) | (226) | (235) | (244) | (254) |
| Free Cash Flow | 64.4 | 64.2 | 66.5 | 66.8 | 69.5 | 72.3 | 75.1 | 78.2 | 81.3 |

Table 7.3: Defining four scenarios for Claringdon Glassworks

| Scenario: | RONIC | Growth rate by year | | | | | | |
|---|---|---|---|---|---|---|---|---|
| | | 2013 | 2014 | 2015 | 2016 | 2017 | 2018 | 2019 |
| 1 | 9% | 4.0% | 4.0% | 4.0% | 4.0% | 4.0% | 4.0% | 4.0% |
| 2 | 10% | 3.0% | 3.0% | 3.0% | 3.0% | 3.0% | 3.0% | 3.0% |
| 3 | 8% | 4.0% | 5.0% | 6.0% | 7.0% | 7.0% | 7.0% | 7.0% |
| 4 | 11% | 1.5% | 1.5% | 1.5% | 1.5% | 1.5% | 1.5% | 1.5% |

Whereas in 2012 it invested a total of 205.2 and realized incremental NOPAT of 5.7 (correcting for a rounding error in the table above), resulting in a RONIC of 11.9%, in 2013 it invested 218.4 to realize additional NOPAT in 2014 of only 4.8, representing a RONIC of only 9%. Because the RONIC of 9% is below Claringdon's OCC of 10%, investments in 2013 are, on average, value-destroying.

What will happen if, in the face of this deteriorating competitive advantage, management decides to ramp up investment with the aim of maintaining the same growth rate as before or, fingers crossed, increasing it? As an alternative, what if, in acknowledgment of the new reality, they scale back growth plans in an attempt to refocus and rebuild competitive advantage?

The answer can go a few different ways. Following are four different scenarios that define the average RONIC the company can earn on its future projects, including a target growth rate:

- In **Scenario 1**, Claringdon grows at 4% per year, slightly lower than the 5% average annual growth it achieved over the past few years. This growth rate, however, is still above that allowed by the company's SCA, assuming that it limits investment to positive-NPV projects. The average RONIC (the average of the positive-NPV and negative-NPV projects which together provide the 4% growth) is 9%. The resulting value of the company is 1,177.
- In **Scenario 2**, management recognizes that Claringdon's competitive situation has changed, however, their adjustment isn't sufficient to rule out some value-destroying investments. It settles on a targeted growth rate of 3%, achieved, unfortunately, through a combination of value-

creating and value-destroying investments such that the average RONIC is equal to 10%. The consequent value of the company is 1,203.

- In **Scenario 3**, management fails to recognize the loss of competitive advantage and senses that the decline in growth is having a negative impact on morale, perhaps because bonuses, promotions, and overall employment fall as profits decline. It elects to focus on growth without regard to value creation. The strategy delivers rising growth rates, from 4% to 5%, then 6%, and finally 7%. Except it requires investment in a growing number of negative-NPV projects. The average RONIC is driven down to 8%, and Claringdon's value declines to 1,115.
- In **Scenario 4**, recognizing the loss of competitive advantage, management eschews ambitious growth targets and instead decides to re-group and re-focus. Based on a careful assessment of the company's competitive situation, growth targets are lowered to 1.5%. The new strategy is successful in re-building and exploiting the competitive advantage, while avoiding tempting value-destroying projects, to the point of obtaining an average RONIC of 11%. Because the OCC is 10%, the average investment creates value such that, even with the lower growth rate, the value of the firm is 1,210 – higher than in any of the other scenarios.

The figures in Table 7.4 below show NOPAT and free cash flow, by year, for each scenario:

Table 7.4: Estimates of NOPAT and free cash flow under each scenario

| | NOPAT | | | | | | |
|---|---|---|---|---|---|---|---|
| Scenario: | 2013 | 2014 | 2015 | 2016 | 2017 | 2018 | CAGR |
| 1 | 120.3 | 125.1 | 130.1 | 135.3 | 140.7 | 146.3 | 4.0% |
| 2 | 120.3 | 123.9 | 127.6 | 131.4 | 135.3 | 139.4 | 3.0% |
| 3 | 120.3 | 125.1 | 131.3 | 139.2 | 148.9 | 159.4 | 5.8% |
| 4 | 120.3 | 122.1 | 123.9 | 125.7 | 127.6 | 129.5 | 1.5% |

| | Free Cash Flow | | | | | | |
|---|---|---|---|---|---|---|---|
| Scenario: | 2013 | 2014 | 2015 | 2016 | 2017 | 2018 | CAGR |
| 1 | 66.8 | 69.5 | 72.3 | 75.1 | 78.2 | 81.3 | 4.0% |
| 2 | 84.2 | 86.7 | 89.3 | 92.0 | 94.7 | 97.6 | 3.0% |
| 3 | 60.1 | 46.9 | 32.8 | 17.4 | 18.6 | 19.9 | −19.8% |
| 4 | 103.9 | 105.4 | 107.0 | 108.6 | 110.2 | 111.9 | 1.5% |

Notice that while Scenario 3 has the highest growth rate in NOPAT (and revenues), it has the lowest level of free cash flow and, ultimately, the lowest overall value because of the need to invest heavily to support a high growth-rate target in the absence of competitive advantage. The next lowest level of free cash flow is Scenario 1, which again suffers from the need to invest heavily because of growth targets that exceed the company's ability to identify and execute positive-NPV investments.

To understand why the investment needs to be higher when the RONIC is lower, think about the following expression:

$$NOPAT \times g = NI \times RONIC$$

where NOPAT is net operating profit after tax, g is the growth rate for NOPAT, NI is Net Investment, or the investment beyond that required for maintenance purposes, and RONIC is the return on new invested capital.

More precisely, NI is the sum of additional investment in the Working Capital Requirement[5] and Capex after deducting depreciation (which is used as a proxy for maintenance investment).[6] Thinking slowly through this expression, we see that the NI needed to deliver growth in profit of *g%* depends on the RONIC to be earned on the new investment. The lower the RONIC, the more investment is required to deliver the desired growth.

Why must this expression be true? Consider a simple case in which NOPAT is currently 100 and we wish to grow it by 5% per year. In other

---

[5]The Working Capital Requirement is the sum of operating cash, trade receivables, inventories, and prepaid expenses, net of the sum of accounts payable, advances from customers, and accrued liabilities such as unpaid taxes and salaries. This investment can be viewed as the additional investment required to get fixed assets (i.e., property, plant, and equipment) to work. For example, a new retail store cannot increase company sales and profits unless it is stocked with inventory.

[6]The idea here is that depreciated assets must be replaced by the company to maintain its physical stock of assets. Therefore, depreciation is a proxy for required capital expenditure. Expenditures above this amount are considered "Net Investment."

words, we must deliver 5 in new NOPAT (NOPAT $\times$ growth $= 100 \times 5$ % $= 5$). If our current competitive advantage provides us with access to projects with an expected RONIC of 10%, the total net investment required is 50, so that earning 10% on the 50 provides the 5 in new NOPAT. In other words, NI $\times$ RONIC $= 50 \times 10\% = 5$. Here's what we mean:

$$NOPAT \times g = (100 \times 5\%) = 5 = (50 \times 10\%) = NI \times RONIC$$

So that,

$$NOPAT \times g = NI \times RONIC$$

Using this formula, and assuming again an OCC of 10%, we can generate a summary table (Table 7.5 below) to help visualize the relative contributions of growth and RONIC to *value*:

Table 7.5: Summary table showing relative contributions of growth and RONIC to *value*

|  |  | RONIC | | | | |
|---|---|---|---|---|---|---|
|  |  | **7.5%** | **10%** | **12.5%** | **15%** | **17.5%** |
| Growth rate, | −2% | 1,056 | 1,000 | 967 | 944 | 929 |
| g | 0% | 1,000 | 1,000 | 1,000 | 1,000 | 1,000 |
|  | 2% | 917 | 1,000 | 1,050 | 1,083 | 1,107 |
|  | 4% | 778 | 1,000 | 1,133 | 1,222 | 1,286 |
|  | 6% | 500 | 1,000 | 1,300 | 1,500 | 1,643 |

Table 7.5 reveals that:

- If RONIC is less than the OCC, growth destroys value.
- If RONIC is greater than the OCC, growth creates value.
- If RONIC equals the OCC, growth has a neutral impact on value, because the average of all future investments is zero-NPV.
- If growth equals zero, there is no investment (beyond maintenance levels) and therefore no return on new investment, so the impact on value is nil.

A different way to think about it is this: when growth targets are disconnected from true value creation, growth itself may lead either to value creation or value destruction, but usually the latter. When a target is focused on growth for its own sake, whether in the form of revenues, operating profits, or market share, delivering on the target often requires value destruction for the simple reason that access to value-creating growth is strictly limited by the nature and extent of the company's SCA.

The human factor plays a key role in this. Sensing the importance of delivering on the growth target to keep their jobs, managers will fight to deliver it, even while many of them will have an uncomfortable feeling in their gut that what they are doing is quite a distance from the right thing to do. But as we've noted previously, it isn't an easy situation, and you may be asking yourself right now what you would do, and perhaps coming up with an answer you're not impressed with. Don't beat yourself up; it's human nature. You need to take care of your tribe.

It's the system that needs fixing, not you. If your only chance to get a nice bonus and a promotion depends on hitting the target, and your family's well-being depends on you getting that nice bonus and promotion, there you have it. One might romantically hope that the resulting value destruction in each such case would create a certain collective resistance to growth, slowing down further value destruction and preventing the ultimate demise of the company. But as we all know, this is not what happens in reality, because each of us accomplishes what we can within the circumstances, or the system, or the culture, in which we're placed. This is the reason it is so easy to find evidence of companies that set growth targets and then aggressively pursue them until they destroy the firm, or at least until painful restructuring is needed.

Notice too, from Table 7.5, that if the RONIC on future investments is only 7.5%, and you insist on growing by 6%, value will be halved, from 1,000 to 500. Now imagine that instead of investing in a series of negative-NPV projects, you convert the company's assets into lottery tickets. What would be the value of the expected cash flows if you were to invest 1,000 in the lottery? The typical (legal) lottery pays out about half of the funds collected, which means that the expected cash flows from a 1,000 investment in lottery tickets, offer a value of 500. Simply put, investing to achieve 6% growth when your investments are expected to earn 7.5%, and the OCC is 10%, will have the same impact on the value of the firm as if you were to convert all of your assets into lottery tickets. You might get lucky, of course, but is that any way to run a business?

## Value-Destroying Growth: A Cautionary Tale

In the 1990s, while the global oil industry was consolidating and facing record low prices for petroleum, one major player found it hard to keep up because of its unique ownership structure. Nevertheless, the company sought to retain its industry ranking for a number of reasons, some value based, like economies of scale and scope or negotiating power; others decidedly separate from value, like ego and vanity. To keep pace with its rivals, the company set ambitious organic growth targets at a time when oil prices were low. The effect, of course, was to make value destruction a virtual certainty.

The intended growth took hold and continued. After several years of this progress, the newly hired head of strategy decided to analyze performance for a large number of projects from a blue line perspective. That is, she decided it would be a good idea to assess outcomes of this initiative relative to expectations at the time of the investment decisions.

It will not surprise you to hear that what she found was deeply distressing. In nearly all cases, the delivered performance was demonstrably below that of expectations, at least the expectations reported in capital budgeting documents. Because the company had been using rigorous NPV planning tools for years, and its senior managers had a well-deserved reputation for being among the most competent and knowledgeable in the industry, the head of strategy suspected that the systematic error could be explained by only one factor: a lack of honesty in the projections. In other words, the reported expectations of future cash flows far exceeded the true expectations held by the managers requesting the capital.

When the head of strategy confronted the offending parties with her suspicions, several managers admitted that they knew there were simply not enough positive-NPV projects to deliver on the company's growth targets. But the targets had to be met, so they re-worked the assumptions on negative-NPV projects, disguising them as positive. The dual outcome was the same one we've observed in too many red line companies to count. The targets were hit . . . and billions of dollars in value were destroyed.[7]

## If Growth is So Bad, Why are Companies Obsessed with It?

Marketing consultants Hermann Simon, Frank Bilstein, and Frank Luby claim that what they call the "market share movement" can be traced back to two main intellectual influences.[8] First, the "PIMS study," a famous research report published in 1987 that reported a strong correlation between market share and profit margin.[9] The market leader in a given industry was found to enjoy profits (measured by pre-tax return on investment) that were roughly three times greater than the fifth-largest competitor. An obvious conclusion drawn from such a finding is that if you want more profits, you need to grow, and the faster, the better.

The second important source for this growth obsession is the so-called "experience curve." As Simon et al. explain, *This concept says that a company's cost position depends on its relative market share. . . The higher this [position], the lower that company's unit costs should be. The market leader automatically has the lowest costs in the market and therefore the highest profit margin.*[10] The implication is that companies with higher market share enjoy economies of scale and pricing power not available

---

[7]Senior managers from the company revealed the estimated losses to us, but asked us not to reveal the figure.

[8]H. Simon, F. Bilstein, and F. Luby, *Manage for Profit, Not Market Share*, (Boston: Harvard Business School, 2006), pp. 8–10.

[9]PIMS stands for "Profit Impact of Market Strategy," and was designed to determine, based on rigorous empirical testing, which business strategies lead to success in particular industries. The data were drawn from thousands of business units in hundreds of companies. The "PIMS study" cited by Simon *et al.* is *The PIMS Principles: Linking Strategy to Performance*, by R.D. Buzzell and B.T. Gale, (New York: Free Press, 1987), and was based on data gathered in the 1970s and early 1980s.

[10]Simon *et al.*, *Manage for Profit, Not Market Share*, (Boston: Harvard Business School, 2006), p. 9.

to their smaller competitors. Again, the apparent lesson for corporate leaders would seem to be to drive market share as high as possible and leave it at that. In the 1970s, the Boston Consulting Group became one of the main promoters of this idea.

Subsequent research has chipped away at both concepts, but the growth fixation has never lost its appeal at the highest rungs of the corporate ladder. The dot-com boom of the late 1990s contributed mightily to sustaining the illusion, with loss-making startups justifying their existence to investors and stock market analysts on the basis of contrived metrics such as "eyeballs" or "website page views," while distracting everyone with assertions that profits were "old economy" and the like.

Simon *et al.* believe another reason for the growth mania is that *"Business schools initiated thousands of MBA students into the market-share cult. Those who earned their MBA degrees in the 1970s and 1980s – and who soaked up the philosophy in its freshest, most concentrated form – now hold C-level positions."*[11] Indeed, one of our motives in writing this book is to conquer the market share obsession and replace it with – you guessed it – the obsession to create value.

While the factors identified by Simon *et al.* have certainly played a part in promoting value-destroying growth, other factors may be more important still. For example, an additional reason for the staying power of the "grow at all costs" philosophy is the way in which popular measures associated with business growth, such as sales volume, revenue growth, and market share, are indeed excellent indicators of whether a company has created SCA through innovation. After all, successful innovators, all else being equal, deliver greater volume and revenue growth than their competitors.

The problem is that managers too often interpret cause and effect backwards, assuming that if the company achieves high growth and increases market share, it must mean that it was "successful," when in fact the causality is frequently in the opposite direction. Genuine innovation, and the competitive advantage that logically follows from it, translate *into* growth. We should no more assume that high growth translates into

---

[11] Ibid., p. 14.

innovation or success than we should assume that the red line drives the blue line. Too many firms have been managed into oblivion because senior management failed to understand this critical distinction.[12]

## The Problem with Executive Pay

All of the reasons above are significant contributors to the obsession with growth and the value destruction that so often results. But perhaps the *most* significant reason for the persistence of growth-driven strategies is the compensation many senior executives receive for delivering on them. The forces that led to the widespread adoption of value-destroying pay policies have been decades in the making. You may be surprised to hear that executive pay practices, at least in the US, were often more value driven in the first half of the twentieth century than they are now.

In 1922, General Motors adopted a bonus plan that awarded management 10% of profit in excess of a 7% return on capital. This approach was a simpler (and, as we will see later, better) version of the EVA-based compensation practices that became popular in the 1990s. Since value is a function of a company's ability to generate returns above the cost of capital, paying bonuses directly linked to how much profit the company earns above this threshold is an effective way to align management and shareholder interests. In effect, the GM plan was a comprehensive bargain between management and shareholders, covering all incentive compensation, both cash and equity-based. Moreover, given their long-term tenure, senior managers had little incentive to boost earnings in the short term at the expense of long-term value creation.

---

[12]Chan Kim and Renée Mauborgne point out another serious problem with this market-share obsession. It tends to define markets by reference to existing customers, leading to increasingly "bloody" competition with industry rivals ("red-ocean strategies"). Instead, managers are urged to focus on "blue-ocean strategies" that consider the potential profitibility from targeting "non-customers" with products or services that make existing competitors irrelevant. This idea is the subject of their bestselling book, *Blue Ocean Strategy*, (Boston: Harvard Business Review, 2005).

The outcome? GM in the early- and mid-twentieth century never fell victim to the problems that arose in Daimler-Benz (and Daimler-Chrysler) under Jürgen Schrempp's economic profit scheme discussed earlier. Bonuses weren't linked to specific targets, and managers expected to hold their jobs for many years, so what incentive did they have to maximize short-term profits at the expense of long-term competitiveness?

This type of approach wasn't exclusive to GM at the time. Such formulae were common among large American companies before World War II.[13] Sadly, comprehensive incentive formulas started to wane in the 1950s, due primarily to two factors. First, changes in tax laws encouraged more companies to grant stock options, but almost no companies charged option values to the bonus pool.[14] They created a separate bonus pool instead. On its own, the growing use of options was not problematic. If used more judiciously in fact, it might have strengthened wealth-creating incentives, but because there were now two separate formulas, one for cash bonuses and the other for equity incentives, companies gradually abandoned the discipline that came with a single comprehensive formula.

Under the old GM system, senior managers received bonuses only to the extent that the cost of capital had already been covered. In addition, there was a single sharing percentage, so everyone knew management's cut of the value created. The system was simple, tidy, and fair. However, the GM approach of giving managers a percentage of economic profit was compromised by the widespread adoption of stock option programs because the value of the option grants was not directly linked to economic profit earned. It could have been, but few companies bothered to make the effort. A significant portion of management rewards had been disassociated from value creation.

---

[13]A 1936 study by future Harvard Business School dean John Baker found that 18 of 22 companies analyzed gave management a share of economic profit or profit above a specified dollar threshold.

[14]Also, option pricing models such as Black–Scholes were not available before the early 1970s. The lack of a credible pricing model would have rendered any attempt to charge a single bonus pool with the value of option grants problematic at best.

The second, and more direct, reason for the gradual abandonment of comprehensive incentive formulas, was the new focus on "job value" and "competitive pay" and the correspondingly lessened attention to an individual manager's contribution to value. The human resources movement arrived in full force after World War II, fundamentally altering the way everyone in business would be paid, including those at the very top. In the US, the first American Management Association survey of executive compensation was conducted in 1950 and the Hay Guide Chart for job evaluation was standardized the following year. The early compensation surveys used profit as a measure of size, soon switching to revenue. Henceforth, benchmark pay would be defined in terms of revenue, not profit or value creation.

Meanwhile, over the ensuing years a large industry of compensation consultants emerged, with the benchmarking of pay serving as the "bread and butter" of their practices. Through these consulting firms, compensation practices in the US gradually found their way into European and Asian companies.

IBM provides a good example of the transition to modern human resource practice and its effect on compensation at the senior managerial level. When Thomas Watson joined the Computing Tabulating Recording Company as CEO in 1915, he had a salary and a 5% share of after-tax profits, net of dividends. As late as 1960, IBM had 90 executives with well-defined individual shares of corporate profit. In the mid-1960s, a major consulting study led to a new compensation program that eliminated individual profit shares, introduced formal job evaluations, and established target compensation levels. By the 1970s, the comprehensive shareholder-management bargain embodied in the GM approach of the 1920s was all but finished.

Competitive pay policies gradually became the norm just about everywhere you looked. The practice begins with the identification of a peer group for each management post. Peers are usually found among other companies, including competitors, with adjustments made for size. In most sectors, size is defined in terms of revenues. Imagine a manufacturer of small household appliances that generates revenues of $3 billion. With the help of compensation consultants, a peer group is defined, consisting of CEOs of similar companies having similar revenues. After the consultants determine the distribution of pay among the CEO's peers, a target is then set – for example, fiftieth percentile pay. This means that if about half of the peer-group CEOs make more than $2 million while the other half make less, the target level of total annual compensation for our CEO is $2 million. The consultant will also advise the company, based on the practices of comparable firms, how to divvy up the $2 million among salary, annual bonus, deferred bonus, stock and stock option grants, and pension benefits.

Competitive pay policy is now virtually pervasive. There are a couple of reasons. First, it reduces the risk of attrition by ensuring that pay is roughly comparable to "market," thereby decreasing the chances of losing good people to poaching competitors. Second, competitive pay policy ensures that pay is not too far *above* market, therefore allowing shareholders to sleep well at night. Corporate boards then try to create strong incentives by putting a high percentage of pay at risk. That is, a significant portion of pay in any year is tied to performance, with performance most often defined as some measure of profit.

There are serious flaws to this practice. First, percent of pay at risk is not a good measure of incentive strength, a point we'll return to in a moment. Most of an executive's pay package can be at risk even while financial incentives for wealth creation are weak or non-existent. Second, a competitive position target (for example, target pay at the fiftieth percentile for a manager's peer group) is not a sensible retention objective. It is not needed to retain poor performers, and is usually insufficient to retain superior performers. Worse, by creating a highly "transactional" culture, it may have the opposite effect to that intended.

To illustrate another problem with competitive pay, let's return to our $3 billion appliance company. The CEO learns that while her target pay is $2 million, chief executives of competitors with $7 billion in revenue have target pay of $3 million. If the CEO cares about her pay, and let's face it, she wouldn't be a CEO if she didn't, what steps can she take to increase it? Well, she could always try to create plenty of value and hope to get a cut. That's the preferred approach, naturally, but then there is the thorny problem that value creation is hard.

Fortunately for our wealth-seeking CEO, there is another option: growth, and lots of it. If she can grow revenues from $3 billion to $7 billion, perhaps by making several acquisitions, future benchmarking exercises will compare her to a different peer group. Now we're talking.

Instead of comparison with CEOs of $3 billion companies, her pay will be benchmarked to CEOs of $7 billion companies, who get paid 50% more. It's a solution beautiful in its simplicity. But, oops, there is the slight issue that she won't pay any mind to whether the investments she makes are positive- or negative-NPV. A boatload of trouble awaits.

The near pervasiveness of competitive pay policies explain why even pay systems based on EVA or economic profit usually fail to deliver the promised results. What good is linking bonuses to EVA (profits above the cost of capital) if the targets are recalibrated each year to ensure competitive pay in the *following* year? Since market pay estimates are largely unaffected by company performance, annual recalibration to competitive pay levels creates systematic and bogus "performance penalties," and the red line spiral takes hold.

These benchmarking and recalibration practices also explain why the percentage of pay-at-risk says almost nothing about the effectiveness of wealth-creating incentives for senior managers. Our fictional appliance company CEO might have 80% or more of her total compensation at risk – a typical percentage among CEOs of publicly traded companies – yet her actions can still destroy huge amounts of wealth.

Unless things start to shift toward the blue line.

# Chapter Eight

# Creating a Blue Line Culture

Returning to our original premise, humans see value in the form of happiness, and the businesses whose job it is to deliver that happiness see value – if they are looking at it properly – as *the expected future free cash flows discounted at the opportunity cost of capital*. In other words, in both cases the principal goal is to allocate energy to pursuits where the expected return more than compensates for the opportunity cost of using that energy. Doing so leads always to survival; not doing so leads always to death.

You know by now that the former behavior represents blue line thinking and the latter red line thinking, and we have hopefully convinced you that companies which uphold the Blue Line Imperative thrive where others inevitably fall by the wayside. The problem you have no doubt noted in previous chapters, perhaps multiple times, is that no one person can force a blue line culture on everyone else in an organization. How, you may be asking yourself, am I to execute a blue line perspective in a red line company?

The effort is indeed uphill for many companies in which the red line approach seems to have become entrenched. But just as people are capable of significant change, so too are businesses, whether they employ five people or ten thousand. To promote a blue line culture – that is, to achieve consistent, repeatable value creation, and to survive and thrive into the future – there are certain behaviors the company must embed. Remember that any behavior that uses less energy than the alternatives for the same outcome must be value-creating, and by the same token any behavior that delivers more happiness than the alternatives while using

Figure 8.1: Developing a culture of blue line management

| Evolution:<br>• Created the<br>human mind | Human Mind<br>• Determines the<br>environment | Encourages the<br>behavior | Behavior<br>• Consistent with<br>value creation | Value Creation:<br>• More happiness<br>with less energy |
|---|---|---|---|---|
| The human mind is the product of evolution and it has the capabilities and interests it has for a reason (to use the least energy for maximum gain). | The human mind controls behavior and sustains the environment, but it will only do so if it sees an interest and has an ability to do it. | The environment will drive the right behavior only if it takes into account the properties of the human mind which needs to deliver the behavior. It forms the bridge between what we are and what we are looking for | Behavior will either lead to value creation or value destruction. Design the environment to ensure it is doing the first. | Value in a business context is defined as that which we need to deliver in order to still have a healthy company one hundred years from now. |

Ask *why*, repeatedly, for each of these building blocks

the same amount of energy must also be value-creating (see Figure 8.1 above).

So how do we get behaviors that are consistent with value creation while avoiding those that aren't? It comes down to people and their environment. We humans, for all our sophistication, remain fairly simple beings. We assess the environment in which we're placed, and then we do things to adapt to that environment. We've done this since the beginning of our existence and will continue to do so until the end of it.

Multiple branches of scientific investigation have shown us that, if you wish to elicit a certain behavior, you need to adjust the environment. Business is no different. If the environment is aligned with the behaviors that promote value creation, we will observe those behaviors. If the environment is inconsistent with the desired behavior, the result will be something else.

All businesses seeking to produce a blue line culture must understand that having to work with humans is an advantage. Our minds are products of evolution, designed to function in a particular way, in pursuit of a particular objective. The main hurdle to value creation isn't that people lack awareness of the difference between the blue line and the red; it is, as we've pointed out, that they often find it to be in their immediate interests to destroy value. Change the incentives and you'll change the behaviors. Organizing and developing a value creation culture is eminently possible. You just need to create the conditions for it to happen.

Figure 8.2: The six sub-sections of the human brain

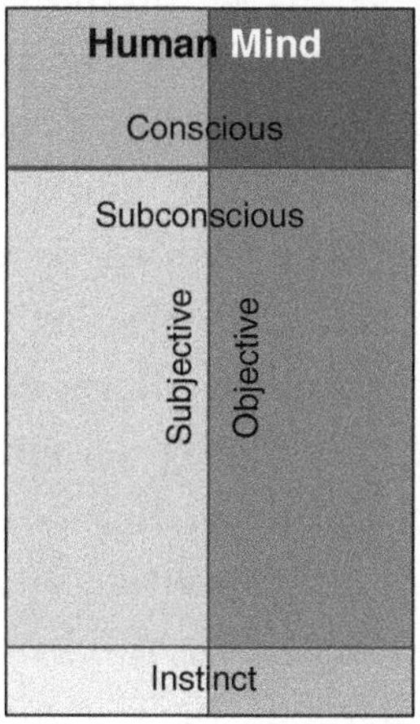

In Figure 8.2 above a simple breakdown of the human mind, consisting of six distinct areas – the three levels we describe as Conscious, Subconscious, and Instinct, each split between subjective and objective thinking. We can think of conscious-level thought as "software-based" processing that is energy-intensive and slow. We say "software-based" because, at this level of processing, new coding is required by the brain to develop a solution for a previously unknown problem. A new situation has presented itself. The subconscious has no previous data available for dealing with it, so new information must be created and codified for use the next time.

Think of it this way. Hiking through the forest, you suddenly encounter a bear for the first time. Your brain naturally raises the question, *"What should I do?"* but, having no previous data with regard to these particular circumstances, it does not have an answer. You haven't seen a bear in the wild before. You need to develop a solution to tackle the question, and quick.

Two different processes will be involved in forming this solution. One is an objective assessment of the situation, combining available data and mathematical calculations of probabilities for various potential outcomes. The other is a subjective assessment in which, having taken this objective information into account, you will then decide whether it is in your best interests to reason with the bear, make friends with it, puff your chest out to make yourself look as big as possible, or run like hell. It is like writing software code that tries to use the processing capabilities of a computer to answer a given question. The process is highly intensive and doesn't proceed very quickly. Asking your conscious brain to answer a question it has never dealt with before is no different from asking a computer chip to draw a circle on a display for the first time.

In contrast, subconscious-level thinking is "hardware-based" processing, and it boasts the virtue of being astoundingly energy-efficient. As the brain is exposed to the same, or highly similar, questions time and again, it begins to embed a kind of muscle memory – automatic solutions that use past learning to address current circumstances with great speed and little energy. This ability has been critical for the survival of our species in an intensively competitive evolutionary environment. Without it, none of us would be here.

When our human brains encounter a certain set of circumstances and need to answer a question regarding how to manage them, both the answer and its related outcome immediately furnish that brain with learning that can be applied to similar future encounters. If we stick around with the bear and the result is getting mauled, we will choose to answer the question differently next time; that is, our brain will reject the earlier answer, since it didn't lead to much success. But if we choose to turn tail and run, and because of this choice we escape unscathed, we will automatically use the same answer, or solution, when faced with the same question.

With each new experience, these solution mechanisms are tested and refined, until they become encoded within the subconscious. Meet enough bears and your solution response will become automatic. Your brain will know to bypass the conscious step altogether and run away every time. The subconscious level is analogous to having a processor with ready-made answers to common questions; the conscious level requires new coding. The human mind is the product of many millions of years of evolution, during which it evolved in a natural way to support the continuation of the species. One of the developmental features of this natural evolution is vast, highly efficient processing capability.

That is, it's highly efficient at the subconscious level, where things happen effortlessly. The processing capability available to our conscious existence is much less efficient. Our fast-thinking, energy-efficient sub-conscious is designed to handle extraordinarily complex questions with great speed to ensure that we aren't eaten by other organisms. To fulfill its evolutionary role, the human brain has devised particular processing capabilities that help us answer the questions most relevant to our con-tinuation, such as *"Do I like eating this kind of berry?"* or *"Do I like sharing physical space or time (a cave during the same night) with a hungry lion?"* or, more important, *"How do I throw this rock or spear to increase the likelihood that I will kill that mastodon instead of just scaring it away?"*

The math required to answer these questions is, naturally, complex in the extreme. Using a more modern example, could a professional basketball player work out the differential equations needed to score a basket? Almost certainly not. However, at the subconscious level, his processing capabilities are more than up to the task, since this is the exact type of complex question our subconscious brain has been designed to answer.

Beneath both the conscious and subconscious levels of our brains is instinct, another "hardware-based" form of processing and the most energy-efficient of all solution processes. Instincts get passed down through generations so that newborns have them hardwired into their brains without the need for learning them via experience. They have the solutions ready the first time the questions appear. Baby marsupials automatically find their way to their mother's pouch. Newborn mammals know to suckle their mother's breast. Human infants instinctively grip onto their father's finger. The answers are there, inside, ready and waiting.

An understanding of the role of instinct is profoundly important to the Blue Line Imperative, and the primary reason is because our percep-tion of fairness is instinctive. Through millions of years of evolution, all humans have developed an instinct for value creation. If we had not, we could never have been so successful as a species.

To illustrate this idea, we ask participants in our executive seminars, *"How many of you have witnessed, or perhaps been involved in, value-destroying behavior that was done in order to deliver on a particular indicator target?"* These seminars usually consist of between 20 and 40 participants from different nationalities. In response to this question, 80% to 100% of the participants will raise their hands.

When we then ask, *"How did you know the behavior was value-destroying?"* we begin to observe the role of instinct in blue line management. Virtually everyone, regardless of background or country of origin, whether they work in government, a bank, a non-profit, a family-owned, or a publicly traded company, will share remarkably similar stories, with the same underlying conclusion: that deliberately choosing to use more energy to accomplish a task or an objective that could have been accomplished using less energy will be perceived by all people as value destruction. Some of them may rationalize the behaviors they have observed or themselves done, in order to believe that it was right, but it is always clear that deep down they know value was destroyed. This is not a surprise. The dominant driver of evolution is the efficiency with which any species makes use of energy. When one of us humans, or a group of us, isn't using that energy well, others instinctively understand that we aren't.

There exists a similar instinctive connection between value creation and fairness. All humans, independent of education or culture, will consider it "unfair" when a particular allocation of energy is not associated with value creation. In other words, we sense when the happiness–energy equation is out of line. We have tested this one repeatedly in the executive classroom too. If a promotion or bonus is given to someone because of who their parents are, the school they went to, their mother tongue, how tall they are, how well they lie, their skin color, their gender, or anything else unconnected to their contribution to the success of the organization, the humans who witness this reward will note that it is "unfair."

In trying to encourage and embed a blue line culture, one needs to be aware of these different mental processes and their connection to evolution, and thus to value creation. We all want to create value because we instinctively, and rightly, associate value creation – using less energy to get more – with our own survival. In business terms, we associate value creation with our own individual survival and that of the organization. To orient behavior around any contrasting goal, such as increasing operating margins, attaining market share, or any other red line objective, and to promote people on the basis of delivery toward such objectives, is to demoralize and alienate others in the organization, even if those others may not be able to explain exactly why they are experiencing these reactions. A voice deep inside is telling them that what is happening is bad, and they recognize that this voice is correct.

Whether we are processing solutions at a conscious, subconscious, or instinctive level, we are also simultaneously processing the objective (*"The world around me"*) versus the subjective (*"What does this mean for me?"*). At the conscious level, where we are aware of our thinking and able to reflect on and explain it, we can separate the subjective from the objective when the two strands are in conflict. (That we *can,* doesn't mean we reliably *will.* The point is only that we have the ability to do so.)

At the subconscious level of problem-solving, we get into trouble, because at this level it is far more difficult to separate an objective assessment of the given situation from the subjective. For thought processes occurring at the instinctual level, the boundary between objective and subjective is so blurred that it is practically impossible for most human minds to have any possibility of separating one from the other.

What sort of methods might help us separate the objective thinking that is so fundamental to value-based decision-making from the subjective thinking that, more often than not, leads to value-destroying behaviors? The most basic danger of subjective thought is that it suits the individual doing the thinking but not necessarily the interests of anyone else, or of the organization at large. When the behavior is being driven by subconscious-level thinking, the person in question will find it extremely difficult to separate the subjective from the objective, and thus may argue that a particular decision – launching a new product, expanding a business-unit, laying off employees, etc. – is right or wrong without seeing that they are allowing a built-in bias to drive the viewpoint. Being aware of such bias is sometimes impossible, and for this reason the Blue Line Imperative must instead be governed by a decision-making process that benefits from the energy-efficiency of the subconscious, along with its phenomenal processing power and data management capabilities, but which strips out all subjectivity. We must agree, then, that anything which cannot be explained at a conscious level – and by this we include feelings, intuition, opinions, and beliefs – must not be used in a decision-making context.

There's a word-association game we like to play with the executives in our classroom. We ask them which of a list of certain words are associated more strongly with data and logic and which are associated more strongly with identity and conviction. The first word we propose is "Hypotheses." Typically, the seminar participants state that while they like their own hypotheses, they aren't "married" to them. If data should

## Figure 8.3: Subjectivity conflicts with being data-driven

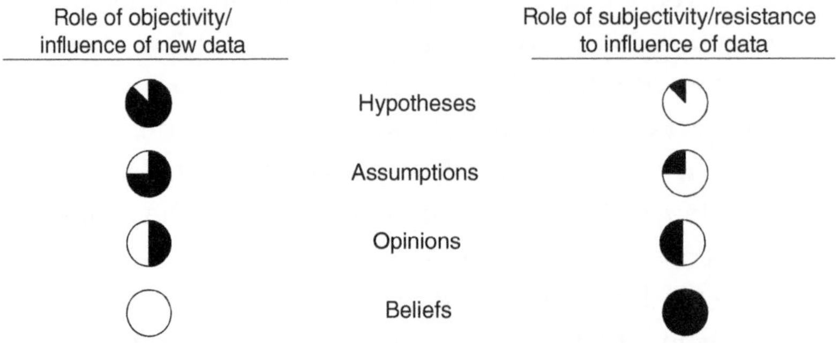

| Role of objectivity/<br>influence of new data | | Role of subjectivity/resistance<br>to influence of data |
|:---:|:---:|:---:|
| ◖ | Hypotheses | ◖ |
| ◔ | Assumptions | ◔ |
| ◐ | Opinions | ◐ |
| ○ | Beliefs | ● |

arise that contradict their hypotheses, they are open to abandoning them and moving on to form new ones.

In Figure 8.3 above, we represent this association with a nearly full moon beside Hypotheses in the "Role of objectivity" column, and a nearly empty moon in the column under "Role of subjectivity". Let's clarify what we mean by the two headings: by objectivity we mean the influence of data and logic. Subjectivity implies the influence of identity and convictions.

It is well known that no two people see the same "truth" in any series of data, since subjective biases will interfere with their mental processing of the data. They will be unconsciously selective about the patterns they pay attention to and those they ignore, and their internal biases will guide the manner in which they explain why the data is the way it is. These biases characterize each person's identity and the convictions they have chosen to honor.

When we learn the scientific method, we are taught to place all personal bias aside and focus only on the data and what it can teach us. When we make hypotheses about the implications of the data for future decisions, we extend the data into the future using "logic." In this context, this is simply a set of rules for processing and extending the information in such a way that even people with very different convictions will agree on the underlying rules, even if their biases prevent them from applying the rules of logic in exactly the same way.

Next we ask our participants to consider the word "Assumptions." Most people perceive this word to be a stronger part of one's identity than "Hypotheses," and therefore less easily susceptible to the influence of data and logic, whether contradictory or not.

This resistance to the influence of data and logic increases as we move through the word list, to "Opinions" and finally, "Beliefs," which are virtually impervious to the persuasions of data and logic.

The universe of business is not the same as the world in which we, as a species, have evolved; it moves faster. In an environment that changes with comparative slowness – our Earth, in other words – subconscious-level thought is highly reliable. If enough of us build the muscle-memory reflex to run away from bears over time, we have a better chance of surviving as a species since fewer of us are going to end up accidentally mauled to death. But in a rapidly changing environment – the world of business – this kind of thought is considerably less useful. In such an environment we must direct the vast processing power of our subconscious in a conscious way, since the solution behaviors we've burned in will become useless quickly, like a high-octane computer that is state-of-the-art when purchased but nearly obsolete three years later due to rapid advances in technology. For learning to occur at the level of the organization, and for a blue line culture to properly take root, we need to be able to explain why we make the decisions we make and examine whether they are truly value oriented.

When we refer to feelings, we are referring to subconscious-level thoughts. Statements such as *"I don't feel I can trust this person"* or *"I don't feel safe in this situation"* are, of course, real and genuine, however they are still underlied by data and logic. What separates these "feelings" from "thoughts" is that the former is processed at the subconscious level and the latter at the conscious. In both cases we are assimilating the data we've gathered about a person, situation, event, or circumstance. But "feelings" occur to us based on questions and corresponding answers we've embedded over time – people who behave a certain way make us feel uneasy – while "thoughts" represent our consciously coding such answers for the first time.

The role of coaching in modern business is to help individuals lower their threshold of conscious-level thought in order to convert "feelings" into "thoughts," that is, to recognize the data influencing the "feeling" statement so that it can be examined objectively. If you can explain why you have a "feeling" about a certain decision, you can then begin to act in a more purposeful, value-oriented way. This is critical to the blue line, and it is for this reason that coaching has become increasingly important for value-oriented managers, or those in blue line companies, in recent decades. The more people in your organization who can tap into their subconscious-level capabilities by becoming more deliberately conscious of them, the closer to constant value orientation you will become.

In order to encourage this process, a blue line culture, while valuing people, does not value intuitions – "feelings." It values the drivers behind those feelings and the learning that can result from bringing those drivers to the surface. To put it plainly, when someone recognizes that their intuition isn't valued, they must shift from subconscious level "feelings" to conscious-level thought, which leads to learning. When, as a result of many people in the company doing this, learning reaches a critical mass, the organization creates value for itself and moves forward.

When we consider any set of circumstances, for example, having to shoot a basketball at a net, we parallel-process both an objective assessment of the action (*"How do I throw the basketball so it goes in the net?"*), and a subjective assessment of whether the action is conducive to survival (*"Is it good for me if the basketball goes into the basket?"*). Any solution we form to address these circumstances – in this case, shooting the ball through the basket – represents the combination of our objective assessment of the attributes of the action and the subjective assessment of how those fit together with our survival.

At the subconscious level, these two cannot be separated, which is why the answer to any question, when carried out only within the subconscious brain, reflects bias. Only when we pull this dual thought process up to the conscious level do we acquire the potential to tease out the underlying data and logic and apply it in an unbiased way. The Blue Line Imperative insists that we tap into the vast store of historical data and processing potential that resides in our subconscious brain and then use it for learning and progress. Red line management does not make this demand. It is satisfied with placing observable, easily measurable targets in front of a human that can then be managed consciously. Feelings remain feelings, intuition and bias are left to run rampant, and no learning is forced. A blue line approach constantly strips down subconscious-level assessments, forces them up, examines where they came from in the first place, and turns the investigation into learning, which becomes value.

Imagine a new restaurant in which you, as manager, are trying to anticipate how many steak dinners will be served each day so that you can determine how many steaks need to be delivered each morning. How would you ascertain the accurate number? Would you ask each of your children to pick a number between 1 and 10 and then average them? This would certainly yield a number, but how relevant would it be to the question being asked? Not relevant at all. You might ask the opinions of your four newly-hired employees, each of whom has worked in restaurants before, and average their answers instead. Clearly their opinions would carry more weight than those of your kids, who have never worked a day of their young lives in the pressure-cooker of a restaurant. Then again, two of the four new employees have worked at vegetarian restaurants in the past, and are themselves vegetarians, whereas the other two worked in kitchens that served a high proportion of meat. Does this make their opinions more relevant or less relevant than the other two employees? It's hard to say. At the end of the day, maybe you should simply take the answer of the person who seems most convinced of his own view.

After some further thinking, you decide to take a different approach. You gather data from market-research companies on the average number of steaks sold per 100 seats in a typical restaurant in your region. This estimate is 40 steaks per night. Perhaps that number, 40, is the obvious one to go with. You share this information with your four new employees and again ask their opinions. All four give different answers than their original ones, each in the ballpark of 40 steaks per day. It seems reasonable and you're inclined to calculate the average of their answers and use it as your daily delivery number.

But you pause a moment by the interruption of another question in the back of your mind: what if you're wrong? What if the number of steaks demanded by your customers is more or less than the number of steaks ordered? If you're off by a little, you could be wasting a few dollars; if you're off by a lot, you could be wasting serious money. Mathematically speaking, if your patrons end up ordering more steaks than the number you ordered on a given day, you'll lose revenue through missed sales opportunities, and who knows how many? Customers may order something else on the menu that evening instead, but the likelihood of them coming back is low. You've squandered not only the margin on the steak for that meal, but you've probably lost all potential future meals from that customer too, and all others they might have told, and the people those people might have in turn told, and so on. If the diners end up ordering less than the number of steaks you ordered for that day, you'll be stuck having to absorb the cost of the extra inventory.

Gathering further data, you confirm that the average number of steaks sold per 100 seats is indeed 40, but you also discover some new information: this number varies considerably on any given evening and can easily be as low as 25 or as high as 60. *Now* how many should you order? Before yet again asking your new employees' opinions, you want to calculate how much having excess steak in inventory costs you versus the carrying cost of the inventory (the Opportunity Cost of Capital (OCC) multiplied by the cost of the steak), factoring in the risk of loss due to spoilage. You estimate this figure to be $1 per steak per day. You also want to calculate, to the extent possible, the amount in lost sales, based on your estimate from industry numbers of the percentage of restaurant clients who do not return for future visits after being informed that their desired menu item is not available. This anticipated lost revenue, adjusted for costs and taxes, comes to $10 per steak per day.

You make the obvious conclusion, based on the data, that you would rather incur the cost of extra inventory than the cost of lost sales. It doesn't take a genius to opt for a $1 loss over a $10 loss, after all. You're aware though, that this strategy is workable only to a certain point, since, as you order more and more steaks, the probability that you will have insufficient inventory to satisfy demand declines, while the probability that you will have excess inventory increases. At the end of all this, you determine the optimum number of steaks to order – that is, the number that will minimize losses – is 48.

When you inform your employees of your decision, one of them says she disagrees. When you ask her why, she responds, *"I think it's too high."* You reply to this statement by asking her what data she has or what calculations she may have performed, and she responds saying, *"I'm just pretty sure it's too high."*

Should you pursue the conversation? On one hand, the employee doesn't have any new data to bring to the discussion, nor a new calculation or analysis of the existing data. Her opinion also doesn't seem to refer in any way to the data and logic behind the number you've arrived at. Her assertion isn't pointing to any potential errors you might have made, either. She's just saying she disagrees, but she can't give a reason as to why.

On the other hand, she is, in a sense, right. That is, despite your fairly rigorous calculation, you can be fairly certain that the number you've estimated will be wrong most nights – the number of steaks ordered will most likely be higher or lower, with the occasional coincidental cases when it is bang on. Her statement that the number seems "wrong" is certainly true, and definitely safe, but in no way helpful. We already knew that the number is wrong. From a blue line perspective, we are not concerned with whether the number is right. We are striving for a methodology that will yield the best possible estimate.

Any decision one makes in business implies a Net Present Value (NPV) calculation, and how delightful it would be to always have access to the full suite of relevant data and the time to perform all of the pertinent calculations. Unfortunately, in the majority of cases, such estimations occur only in the subconscious of the person making the decision, and of the others who judge from the sidelines. Sometimes, this is evidence of a non-value-based company. In other cases, it may simply be evidence that the time and energy required to gather the data and make the necessary calculations are too great to justify the effort. We describe a blue line culture as one that relies on data and logic while eschewing human bias and opinion, but at the same time we must recognize that requiring every decision to be supported by conscious-level evidence is impossible, and therefore, in certain cases, value-destroying.

That said, we can still distinguish a data-driven, value-based process (a blue line environment) from an opinion-driven (red line) one, whether the calculations are captured in a spreadsheet or not. In a red line culture, little or no effort is made to obtain the data in the first place, even for the many cases in which the *value* of gathering and analyzing it would easily cover the *cost* of doing so. For other scenarios in which the value may not be sufficient to cover the cost, it is still possible to make the distinction between red and blue, because the nature of the questions asked to support the decision will always tell the tale.

In a red line environment, where value is not prioritized, common questions used to make decisions might include *"What's in it for me?" "How does this help me get promoted?"* or *"Does this help me hit my scorecard targets for the year?"* Such questions aren't expressed aloud, of course, but you can be sure they are being asked all the time, perhaps at the subconscious level.

One question that does get asked explicitly, and which is even more dangerous than those that might seem more obviously self-serving, is *"Will this encourage the markets to reward us with a higher share price?"* We have discussed in detail why questions of this sort lead companies down the wrong path. By contrast, the Blue Line Imperative guides us toward different questions whose answers will help drive every business decision. Those questions, you now know, include *"How does this impact the value of the business?" "How does this help to drive future cash flow in a way which will cover the upfront investment?"* and *"How does this improve the long-term competitiveness and viability of the company?"*

Conscious-level thought and analysis consume enormous amounts of energy and time in comparison to those at the subconscious level – and can, in fact, be downright painful. The blue line principle tells us that energy must be allocated efficiently and intelligently since it is in limited supply, and therefore whenever the value benefit of analyzing a decision at the conscious level is insufficient to justify the energy and time required, we ought to rely instead on our subconscious processing capabilities. In other words, the great majority of decisions in business, even in organizations that are thoroughly blue line in orientation, will be made without the benefit of Excel. It is important to recognize that doing things this way does not contradict the principles of value-based decision-making. Indeed, given the appropriate circumstances, a value orientation demands it.

The key point of divergence here is not whether the analysis takes place in the conscious (e.g., using spreadsheets or some other formal tool) or the subconscious, but whether it is focused on answering blue line or red line questions. In the final chapter, we provide more examples of blue line questions. We ask these questions in part because, at the end of the day, they help answer the most important question of all in routine, day-to-day decision-making: *"Is this idea a good or bad one?"*

Since we seldom have the time available or access to all the relevant data, few of the business decisions we make can be said to follow a true, or at least comprehensive, value-based methodology. That is, we don't usually have time to plug everything into an Excel spreadsheet populated with lots of fancy formulas that together yield a black-and-white estimation of NPV. Knowing these usual limitations, how should we approach such decisions, many of which we all face in any normal day? To address this dilemma we return to our statement above: we must ask the right questions. Lines of inquiry that progress from a blue line perspective will achieve the desired goal: neutralizing our inherent subjectivity around decision-making.

Though the process will look slightly different depending on the nature of your company, the industry you work in, the size of the organization, the configuration of its staff, and various other factors, there are important steps you can take within your firm to try to encourage blue line thinking and assessment. Here are a few examples:

**Pay attention to the words people use.** In explaining any routine decision you make, you reveal which part of the brain you are using, and the extent to which you are really thinking about the NPV (in the thought process, not in a conscious way, nor in a spreadsheet). The Blue Line Imperative demands an explicit commitment to do away with words or references that hint at personal bias, like *"in my opinion"* or *"I believe that."* Recall our restaurant employee above who simply felt that the number of steaks being ordered was too high, though she couldn't offer an explanation to support this gut feeling.

If you're focused on value, every input should be supported by data and logic. Having a feeling about some decision being right or wrong isn't enough; you need to be able to explain why, or you need to be able to recruit a colleague who can validate your "feeling." In a blue line environment, everyone agrees that simply saying *"I don't agree,"* with no explanation, is counterproductive to the organization's ultimate goal of building value.

It is to the most senior people in the room that the greatest responsibility for upholding this principle must fall. Perhaps the most junior person at the table is allowed to say *"I don't agree"* without having to offer an explanation once, but the most senior person must never be allowed to do so, since it will send the message that others can do the same, and the trickle-down effect will destroy value along the way.

Listen closely to others' explanations behind the decisions they make. Strive to understand when someone is revealing that they are playing fast and loose with the data, or expressing a view that seems more convenient for them or driven by their personal agenda, than oriented around value creation for the company. When people say, *"I think this is a bad idea,"* listen hard. Are they saying it's a bad idea because they stand to lose something, or because it runs counter to the creation of value for the organization?

**Utilize the power of small teams**. Anytime you can benefit from the input of others – for instance, to alert you to the fact that you are employing subjectivity instead of data-driven assumptions – do so. Everyone on a team must feel they are free to comment on and critique the others' input. This philosophy reduces the collective risk of using the subjective subconscious because each member of the team will recognize implicit or spoken assumptions that are different from their own and will thus be able to check the use of erroneous assumptions subconsciously favored by others on the team. Typically, this approach yields the best results with small teams. Why small teams and not big? Because most day-to-day business decisions are small, relatively speaking. Another way to think about it is that the size of the team depends upon the importance of the decision. In our earlier example of the Daimler acquisition of Chrysler, the decision called for a much bigger team than the one used – and the narrow aggregate point of view cost the company dearly.

**Use rigorous, purpose-built methodologies for decision-making.** Processes like Six Sigma are designed specifically as a means of helping people stick to data when making decisions. The Blue Line Imperative has its own similar language, oriented entirely around the goal of using data meticulously to ensure value orientation in all decision-making throughout the organization. Using this language is essential in making value-based decisions. Capture the data and pin it down as best you can, removing all opinion and human subjectivity. Place the numbers in their tidy little cells, apply the mathematical lessons we've discussed, and let them reveal their own truths independent of what you or others may think about what they should represent, or the story you might hope they will tell. And when spreadsheets aren't practical or possible – including when the value of using one is insufficient to justify the time and energy required to use it – go back to the first of the three principles, and listen closely to the words people are using.

## The Three Pillars

In broader terms, any value-driven environment is characterized by three critical attributes, which we refer to as the three pillars of the Blue Line Imperative: Fairness, Trust, and Learning.

As we have alluded to previously, people sense things in an organization, and when things are fishy – in other words, when decisions are being made to satisfy personal agendas or senior managers are being judged based solely on whether they are hitting red line indicator targets – it doesn't usually take long for the scent to be picked up by others. This leads to the perception by one, or more often many, that treatment within the company is not equitable, or not rational, or both, and the collective sense of both fairness and trust quickly diminishes.

Without these qualities, people are unwilling to take the risks that necessarily accompany experimentation and learning. And without learning, value creation simply cannot happen. We come into this world ignorant, and we remain ignorant on many dimensions throughout the whole of our existence. Managing for value demands constant probing of our environment to close knowledge gaps and learn what we need to do to manage our businesses more effectively. As we have said, blue line companies are constantly experimenting in the never-ending search for an understanding of their true underlying value drivers. This requires a willingness to accept failure and learn from it. Such experimentation will not take place unless people truly believe that they can trust the organization, and the others within it. The three pillars are interdependent – either a vicious or a virtuous cycle will take hold. It is up to everyone in the firm, especially those at the most senior levels, to decide which.

Let's recall our discussion of the previous two chapters. Who gets promoted in a red line culture dominated by KPI-based rewards? Is it the value creators or the best liars, cheaters, and manipulators of the system? Those who rise to the top in such systems will often be those who not only destroy value along the way, but who also lack the skills to do what's needed when they get there. How fair does such a system seem, and what effect will this perception have on the morale of everyone working within it?

It is no revelation to say that we all have an innate sense of fairness and can recognize when something is unjust, whether it's an undeserved promotion or a short-cut in manufacturing that compromises quality. What is important to remember is that when we see something being done unfairly, it makes us feel bad because we know instinctively that when things are fair, we can create value, but when they are not, value gets destroyed.

Our discussion about the capital markets was also, at bottom, about fairness. In just about any market, you will hear the term "fairly priced" – a direct comparison of fairness relative to value. When we talk about a financial security being fairly priced, what we are really saying is that in expectation, its price equals its value. When markets are perceived to be fair, more investors and entrepreneurs are drawn because they have confidence that the prices they expect to pay or to receive will be "fair."

With growing numbers of economic agents operating under conditions of reasonable openness and transparency, prices today are becoming increasingly fair, resulting in a virtuous cycle in which the capital markets get ever closer to reflecting nature's OCC. This facilitates, in a very healthy way, the analysis and pricing of risk, more efficient allocation of the world's capital resources, and ultimately, more wealth creation and more stuff.

As historian Mark Smith pointed out when discussing the history of the New York and London stock exchanges, *"Both Wall Street and the City of London recognized that markets that were not perceived as 'fair' could not hope to satisfy the demand for capital of modern industrial economies."*[1] "Fair" meant the same thing in the early years of the stock exchanges as it does today: a dependable proxy for value. We are getting closer to this ideal and, as a result, more capital is finding its way to more of the users who can create value with it. At the very outset we made the claim that it is "fairer" when the right people, that is, those

---

[1] B.M. Smith, *A History of the Global Stock Market: From Ancient Rome to Silicon Valley*, (Chicago: University of Chicago Press, 2003), p. 85.

with the good ideas, get access to funding rather than those with merely the best contacts. Value creation requires that resources are allocated fairly, on the basis of the expected cash flows and not relationships (race, social class, gender, etc.), therefore in any value-driven enterprise, rewards must be granted fairly, that is, on the basis of value creation and not something else. We are getting better at this all the time.

Consider two division managers who are equally competent in terms of their raw skill. Let's call them Pierre and Annette. Each is facing an opportunity in which there are two possible outcomes and each happens with equal probability. Specifically, the opportunity is to invest in a project that will realize cash flows of either 50 or 150, each with a probability of 0.5. Just as in the coin toss, there's a 50/50 chance of either outcome. But here's the difference: Pierre negotiates a price of 80 while Annette invests 120 in her project.

Now suppose that Pierre's payoff turns out to be 50 while Annette's payoff is 150. In both cases, the outcome was the result of random events outside of the control of either manager. Which manager created value? If you take the standard view, you'll decide that Annette did the better job and promote her – after all, she got the maximum payoff, didn't she? Pierre gets fired for losing money.

But look a bit more closely. Which of them was investing in a project with a positive expected outcome? Pierre spent 80 in cash and realized cash flows of 50, for a net cash impact of −30. Annette spent 120 in cash and realized cash flows of 150, for a net cash impact of +30. But don't go promoting Annette just yet. Pierre spent 80 for a project with an expected value of 100, whereas Annette spent 120 for a project also with an expected value of 100. Either outcome carries a 50% chance of occurring. If we apply the same simple equation from the coin toss to both managers here, we find that the NPV of Pierre's project was 100 − 80, or 20, while the NPV of Annette's project was 100 − 120, or −20. Annette destroyed value but got lucky; Pierre created value but was unlucky.

Who would be rewarded in your company? When we ask participants in our executive programs, most say Annette. Well, let's think about the signal promoting Annette would send to everyone else in the organization. People who understand the business will have a strong sense that Pierre made a good decision but got unlucky based on extraneous factors, and that Annette made a bad decision that paid off due to dumb luck.

Suppose Pierre continued applying the same approach to his next 1,000 such decisions. How much would his decisions have made or lost for the company? Each involves spending 80 with an expected payoff of 100; and in each case, the outcome is either 50 or 150, for a net payoff of $50 - 80 = -30$ or $150 - 80 = +70$. For the 1,000 trials, the average value created per project would be $[(0.5 \times -30) + (0.5 \times +70)]$, or $+20$. For Annette, the average value result per project would be $[(0.5 \times -70) + (0.5 \times +30)]$, or $-20$.

The problem, of course, is that we won't have 1,000 iterations to sort out the value-creators from the value-destroyers. Where capital investments and major strategic choices are concerned, the available number of observations is usually small enough that we know randomness will play a not-insignificant role in determining outcomes.

So, as people see Annette getting promoted over Pierre, they come to realize that because of the role played by randomness, it is more important to be lucky than good. Pierre did the right thing after all, and got kicked to the curb anyway. The result is a loss of confidence in the process and in the competence of top management to distinguish good ideas which don't turn out well, from bad ideas which do. Again, just think where we'd be if every time a scientist had an experiment go wrong, he or she were shoved to the sidelines. No light bulbs, for one.[2]

The influence of randomness can trick us into thinking someone is a genius – or a fool – even when they're not. Consider the plight of two separate Hollywood executives. Sherry Lansing had a long and illustrious career in the movies, with numerous hits to her name.[3] Her last big job in the industry was as chairman of Paramount. The studio enjoyed a string of hits while she was there, including *Braveheart, Forrest Gump*, and *Titanic*. Not bad on the resumé. But when these smashes were followed by a series of flops, as is the whim of the industry, Lansing promptly lost her job. Luckily, her previous successes made this hit relatively easy to take.

---

[2] As reported in *The Economist's* Schumpeter column (April 14, 2011), Thomas Edison performed 9,000 experiments before coming up with a successful version of the light bulb.

[3] See Leonard Mlodinow, *The Drunkard's Walk: How Randomness Rules our Lives*, (London: Penguin Books, 2008), pp. 15–16.

Columbia Pictures' Mark Canton wasn't so lucky. After a short tenure, he was fired from the top job because he hadn't produced the box-office returns the Board had been expecting when they'd brought him on board. But then a funny thing happened. After he left, several projects that he had personally green-lit struck gold, with just five of them generating over $1.5 billion in ticket sales worldwide.

The lesson? Randomness plays a role at all levels of the organization, from the most junior associate to the top dog, and if you want to uphold the blue line and truly embed a blue line culture, it is important to recognize this and see people for the value orientation they display and the data-driven processes they employ in decision-making, as opposed to merely the raw results attached to their decisions, which are often misleading. Without more insight on what truly drives success in the movie business, we can't really say which of the two executives is Pierre and which is Annette.

Now imagine you are the film executive. How motivated would you be if a string of box-office failures, even if due to pure randomness, results in you getting fired? If indeed Mark Canton had real talent in spotting future hits, as evidenced by five of his projects eventually delivering huge returns, but he was fired before these results became apparent; other executives will have learned that you can't trust those above you to recognize actual talent and distinguish value-creators from those who got lucky. How confident would you be placing your career in the hands of such people?

We are by no means declaring that success in business investing is entirely random. The Blue Line Imperative demands having the right culture and management systems in place as necessary conditions for business survival. Further, it insists that being a value-creator requires a sincere effort to figure out the expected cash flows from proposed investments and act accordingly.

However, the role of randomness cannot be denied. The problem, as Dan Gardner writes, is that *"we see patterns where there are none. We treat random results as if they are meaningful."*[4] Michael Shermer calls it *"patternicity,"* or the tendency to find meaning in both meaningful and

---

[4]Dan Gardner, *Future Babble: Why Expert Predictions are Next to Worthless, and You Can Do Better*, (New York: Dutton, 2011), p. 15.

meaningless data, a tendency inherited from our Stone Age ancestors.[5] Our brains are programmed this way because the survival of our species depended on it. If the rustling we heard in the trees was not a predator, fine – no harm would befall us. But if it *was* a predator and we ignored it, we were doomed to become some other animal's lunch. In today's world, the first kind of error, a false positive, can be just as fatal as the second, a false negative, with the result that we reward Annette and condemn Pierre. If you aspire to be a blue line manager, you must always try to gather, to every extent possible, the relevant information that can allow you to distinguish value-creating projects from value-destroying ones. As a minimum, that means you need to recognize that if the expected value of the project was 100, Pierre's decision was value-enhancing even if the outcome, on the surface, seemed negative. Your task is twofold. First, you must assemble accurate, honest data to estimate the expected cash flows and their probabilities – something we'll come back to a little later – and second, you must be aware of Pierre's attempt to create value.

The more that you and others can do these two things, the stronger a blue line culture you will create. At the other end of the scale, the red line end, is the problem we have noted more than once, where others have no option but to change the way they work as a response to the realization that promotion or compensation is based entirely on out-comes. Once they understand what outcomes on which indicators drive compensation, they will do what they must to feed their families. First, since they're pretty sure management doesn't understand its own busi-ness, they'll take the budgeting and target-setting process into their own hands. Second, they'll manage delivery to ensure that the targets are met – more often than not, in a value-destroying way. It's the classic, and lethal, red line corollary. Employees deliver on their targets, and value is destroyed.

Let's say you're a manager, and it's budget- and target-setting time. You know that if you work at near-maximum capacity and capability, you

---

[5]Michael Shermer, *The Believing Brain: From Ghosts and Gods to Politics and Conspiracies: How We Construct Beliefs and Reinforce Them as Truths*, (New York: Times Books, 2011), p. 5.

will deliver an expected outcome of 100 for the coming year. When your boss asks, *"What can I expect from you this year?"* what are you going to answer?

When we pose this question to executives in the classroom, the responses we hear are typically, 70, 80, 50, or, occasionally for laughs, 10. Humor aside, it's clear from these answers from highly accomplished businesspeople that everyone's inclination is to manage expectations. Not only that – these responses also give a general indication of how clueless people tend to think the other person is. If you believe your boss doesn't understand the business, you can safely offer 50 as an answer, knowing you can easily deliver 60 or more, thereby securing your bonus and positioning yourself for promotion.

This is of course easier in theory than in execution. Any senior manager with experience is well aware of the tendency to sandbag or build in deliberate buffers and can spot it a mile away. So what do they do in anticipation of such guile? They set "stretch" targets, driven from the top.[6]

Imagine these targets are applied in the above scenario. You tell the boss you think you can deliver 80 for the year. The boss promptly replies that a stretch target is being added to your proposed target, so 120 is the new number. After some negotiation, you and the boss ultimately agree on a final target of 100. In essence, the boss's stretch target serves as a correction to your initial deceit, and so you end up where you pretty much should have been anyway.

An interesting feature of such a red line system is that anyone who makes the mistake of telling the truth will learn quickly not to do it again. Everyone instead learns to lie, leading to two outcomes: first, that targets are largely meaningless, and second, that senior management assumes all of their subordinates are liars. This is neither a healthy nor a productive environment for any organization. It stands to reason that if you assume your people are lying and then organize your control systems

---

[6]We do not oppose stretch targets *per se*, just those that are imposed from the top. The bottom-up stretch targets used in Google or Toyota or other high-performance companies are entirely in keeping with a value-based approach to management.

accordingly, they are going to lie to survive within that system. And so on. Everyone, in their work environments and in their personal lives, makes a never-ending series of cost-benefit analyses. If there is a cost to telling the truth but no benefit, you can be sure that even those with the best of intentions will find it difficult to tell the truth.

An additional outcome of a red line climate – that is, one in which lying-correction mechanisms are necessary – is that the targets and the indicators will become increasingly difficult to interpret as people find newer, more creative ways to lie. Given the lack of reliable information on the business, how can we expect senior management to make sound value-creating decisions?

Apart from the practical impossibility of steering resource allocation in ways that make the blue line go up in such a scenario, morale is sure to decline as well. There are two reasons for this. First, people don't like being made to lie. Lie detectors work precisely because, when we do lie, our physiology betrays our conflict: hearts race, adrenaline starts pumping, pulses quicken, and we sweat – responses typical of an animal that finds itself cornered. Some label it stress. We see it as a direct biological indication that inherently we want the world to be fair. If the external environment offers more rewards for lying than for honesty, we will lie, but we won't like it, and at some point we will start resenting the source of this discomfort. In a business enterprise, the logical result is declining morale, and commensurate declines in productivity, loyalty, and dedication to organizational goals.

The second reason for the natural decline in morale is the one we've mentioned previously. When employees see that those who get promoted are not the best value-creators but instead the most effective liars, indifference to corporate goals turns into something much worse: downright animosity toward both the company and its senior leaders. High performers are especially sensitive to value creation, and thus especially sensitive to its absence. Since fairness and value creation are two sides of the same coin, they have a strong sense of fairness, so they are always the first to recognize the lack of it, and will be the first to become de-motivated. They may stick around out of necessity, but their performance will taper off because they realize there is no reward for value-oriented behavior. Either way, the company squanders the talent and drive of its best people.

# Fair Process

Another perspective on the importance of fairness comes from the work of our INSEAD colleague, Ludo Van der Heyden, based on the pioneering work of two other INSEAD colleagues, Chan Kim and Renée Mauborgne.[7] This perspective does not concern outcomes as such, but rather focuses on the fairness of systems and processes. Ludo compared two approaches to management, the first by results, the second by fair process, and concluded that

*"The more one 'manages for results' . . . the more elusive these results may seem to become."*[8]

Ludo argues that we are hardwired for fairness. He cites, for example, our physiological responses to demonstrations of trust. In such situations we release oxytocin, a hormone linked to sensations of pleasure. *"Humans are reciprocal beings,"* he says. *"They respond to 'fairness' with fairness and to 'unfairness' with hostility."*[9] Management that relies solely on results, if it is seen as unfair, will inevitably fall short.

Kim and Mauborgne, writing in the *Harvard Business Review* in 1997, underscored the importance of fairness and trust, stating that *"When employees don't trust managers to make good decisions or to behave with integrity, their motivation is seriously compromised."*[10] We would add to that by saying it is more than a matter of simply not trusting management to make good decisions. There is the further issue that employees don't trust their bosses to determine whether an employee has engaged in value-creating or value-destroying behavior. Everyone feels forced to

---

[7]Kim and Mauborgne's early work in this area includes, "Procedural Justice, Attitudes and Subsidiary Top Management Compliance with Multinational's Corporate Strategic Decisions," *The Academy of Management Journal*, 36(3), 1993, pp. 502–526; and "Procedural Justice and Manager's In-role and Extra-role Behavior, Failed to Understand this Critical Distinction" *Management Science*, 42, April 1996, pp. 499–515.

[8]L. Van der Heyden and T. Limberg, "Why Fairness Matters", *International Commerce Review*, 2007, 7(92–102), p. 94.

[9]Ibid.

[10]W. Chan Kim and Renée Mauborgne, "Fair Process: Managing in the Knowledge Economy", *Harvard Business Review*, July–August 1997.

deliver the indicator targets as commanded, even if achieving those targets requires the destruction of value. The implicit knowledge that they have been forced to destroy value to keep their jobs is obviously demoralizing and de-motivating, and will inevitably lead to the resentment of the bosses who made them do it.

Any decision-making process in a business – from capital allocation to compensation, promotions to firing – can be perceived as fair or unfair independent of how its outcomes are judged. Outcomes seen as fair are frequently rejected if the processes behind them are believed not to be. Kim and Mauborgne describe the case of a woman who felt she never had her "day in court" because the magistrate ruled in her favor when the hearing had barely begun. Vindicated or not, this woman felt extremely frustrated because she hadn't been able to tell her story.

By the same token, as Van der Heyden and Limberg point out, *"People are much more prepared to accept 'unfair' outcomes if they feel that the process that led to these outcomes was fair."*[11] Consider the classic example of two children arguing over who gets the last piece of birthday cake. How do you keep them both happy and stop the birthday party from turning into a riot? If the piece of cake is cut in half, the two halves will almost certainly be unequal, and one of the children will invariably claim that his is the smaller of the two, a supreme injustice. So what do almost all parents instinctively do? They offer a deal: one of the kids gets to cut the cake, and the other gets to choose the first piece. The reaction from almost all children offered this deal is universal. Both, understanding that the process is fair, agree to it. Even if the final outcome looks unfair because one piece will inevitably be bigger than the other, both children will accept it because the process that led to it was equitable.

The above research tells us that, ultimately, for a process to be fair, the leader must be seen not only to manage fairly, but also predictably. And predictability demands transparency. Finally, it requires "authenticity" (the search for truth is always paramount), willingness to change course as new evidence emerges, equal treatment of everyone regardless of rank, and a communication policy that allows people to express them-

---

[11]Van der Heyden and Limberg, op. cit., p. 95.

selves. We would add that any process that is not inherently value based will automatically be seen as unfair independent of the other criteria. If collective value is not seen as the goal, how can any process be considered a fair one?

## The Perils of Opinion

Decisions based on opinions are, no matter how you slice them, patently unfair. Imagine yourself in a room with 24 others. Each person is asked to state his or her opinion on a complex topic, such as how to resolve the conflict between Israelis and Palestinians. You may perceive at a superficial level that some people seem to agree with each other, but as you dive deeper into the issues it will become clear that each person's opinion is unique; no two agree on every aspect of the issue.

We have then, 25 people with essentially 25 different opinions. If someone in the room is right, or if we are going to resolve the conflict at all, how many of the other opinions must be rejected as being flawed in some way? At least 24. In fact, if we put all of humanity in the room, the outcome would be the same. Since no two opinions are exactly the same, just about everybody has to be wrong in some respect.

Recognizing the impossibility of reaching agreement on opinions, how can we determine which way to proceed? Whose opinions will be honored and whose will be ignored? The truth is that resolving differences of opinion requires a hierarchy, in which somebody is able to rule, imposing his or her opinion on everyone else. Is this fair? And who gets to decide?

In the early years of Amazon.com, a software and new product developer named Greg Linden came up with the idea of using the Amazon shopping cart to make recommendations to customers.[12] Inspired by the candy displays and other impulse buys found in grocery store checkout lanes, he thought it might be profitable to personalize impulse buys for Amazon shoppers. As he explains, *"It is as if the rack near the checkout lane peered into your grocery cart and magically rearranged the candy based on what you are buying."*

---

[12] http://glinden.blogspot.com/2006/04/early-amazon-shopping-cart.html

Linden built a test site and began showing it to colleagues. The reaction was largely positive, but a marketing senior vice-president was afraid that the recommendations would divert people from the more important act of checking out. After this senior VP rallied others to his point of view, Linden was told that he was forbidden to work on the idea any further.

The young developer might have stopped there, but didn't. Instead, he prepared the feature for an online test to measure its sales impact. The senior VP was angry, but the test went ahead.

*"The results were clear,"* writes Linden. *"Not only did it win, but the feature won by such a wide margin that not having it live was costing Amazon a noticeable chunk of change. With new urgency, shopping cart recommendations launched."* Today, as you are no doubt aware, this feature has become ubiquitous on the Web.

The moral of the story? Listen to your customers, and not, as Microsoft's Ron Kohavi memorably phrased it, to the HiPPOs (or Highest-Paid Person's Opinions).[13] The opinions of senior executives may have been honed through years of experience, but that doesn't change the fact that they are opinions. In a blue line, value-oriented environment, data, facts, analysis, and continuous learning must rule the day.

Where learning is concerned, the role of experience can be positive or negative. When used to gain genuine insight about the world around us, experience helps managers pose the right questions, gather the right data, and analyze it in rigorous, logic-driven ways. Too often though, experience is used to confirm biases instead. In such cases, any data gathered will appear to strengthen and support personally skewed opinions. People who gain and use experience in this way become increasingly dangerous from a blue line perspective, as they gather more and more data and experience to support their biases, while dismissing any data that may prove contradictory. Anyone with less experience will have less of either type of data – the data supporting the bias or the data refuting it – and thus will have hardly a leg to stand on when challenging the "experienced" person's opinion.

Figure 8.5 below demonstrates the role of experience in decision-making. "Guesses" are strongly influenced by personal biases, and because

---

[13] Videolectures.net/kdd_kohavi_pctce/

Figure 8.4. Learning vs. anti-learning

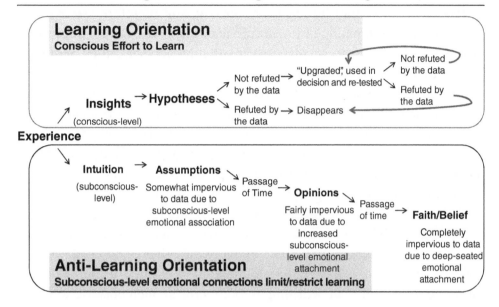

Figure 8.5: The role of experience in learning

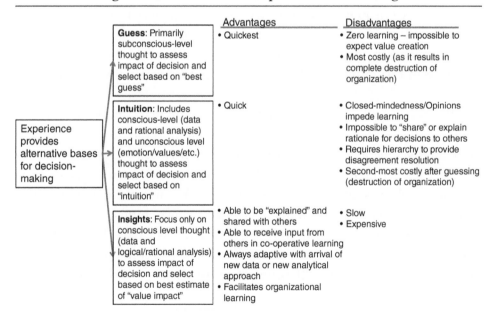

they rely primarily on subconscious thought, can neither be debated nor defended. They cannot be used for learning, and therefore are inappropriate for decision-making in a culture aspiring to the Blue Line Imperative. In addition, because guesses are highly personal, they require a hierarchy to resolve differences.

"Intuition" is potentially less dangerous in its influence on decision-making than guesses, but nonetheless suffers from the presence of personal biases, again requiring a hierarchy for resolution of any differences; because intuition still relies on subconscious thought – though to a lesser degree than guesses – it defies logical explanation to others. The rationale for the intuition may not be known even to the person expressing the intuition, therefore how can it contribute to others' learning?

"Insights" exist, by contrast, purely at the conscious level. They can therefore be explained by the person expressing them, and used not only in data-driven, logic-based decision-making but also as a means to promote organizational learning – both crucial elements of a value-oriented culture.

As Nobel Laureate Daniel Kahneman writes,[14]

> "Most of our judgments and actions are appropriate most of the
> time. As we navigate our lives, we normally allow ourselves to
> be guided by impressions and feelings, and the confidence we
> have in our intuitive beliefs and preferences is usually justified."

In other words, at least according to Kahneman, intuition works – most of the time. No one would argue its importance to our everyday functioning, nor its ubiquity. We develop heuristics, or rules of thumb, to guide us through a practically limitless array of isolated decisions in our personal lives and in our businesses. Is the man we encounter on a stretch of desolate road someone to fear? Since humans first walked upright, those able to accurately assess such risks have had a survival advantage over others, and therefore a higher likelihood of successful reproduction. We are the offspring of those who boasted such skills.

---

[14] Daniel Kahneman, *Thinking, Fast and Slow*, (New York: Farrar, Straus and Giroux, 2011), p. 4.

But intuition can only take you so far, even when the decision-maker's intuition has been honed as finely as the senior executive's business sense. Kahneman cites the example of a chess master who, faced with a challenging position, can rely on intuition to propose several potential moves, but only through purposeful and analytical thought can he identify the *best* one.[15]

Or consider another of Kahneman's examples. The chief investment officer of a large financial institution decides to buy stock in Ford Motor Company. When asked his reasoning, the man says he is deeply impressed with the quality of Ford's cars. His intuition tells him that high-quality cars should translate into better bottom-line results and, presumably, a higher share price.

It is in just such cases that we can see the danger of relying on intuition to guide us in allocating capital or other resources. We easily fall into this trap, substituting an easy question (*"Do I like Ford cars?"*) for the more important, but more difficult, question (*"Are Ford's shares underpriced?"*). As we so vehemently said before, much of what it means to be value driven is asking the right questions. If you don't, you still might get lucky, like Annette, but banking on winning the lottery every time – that is, investing in negative-NPV projects and hoping for fortunate results – is far from a viable business strategy.

We pull this substitution trick on ourselves because intuition works just enough to fool us into thinking we can rely on it even when the situation calls for more careful and reasoned analysis. Also – and this is especially important – intuition imposes little or no strain on our cognitive resources, so it's our preferred default. Gathering data and subjecting it to rigorous analysis are purposeful and deliberate acts. They're hard, complex, and taxing. But they're the only way to take care of the blue line.

Have you ever come away from a cocktail party thinking, *"Wow, that guy I was talking to had really great opinions. I wish I had his opinions."* No, you haven't – because you love your opinions, just as we all do. If you didn't love them, you would have changed them much more often than you have. Your deep emotional and subconscious connection to the

---

[15] Ibid., p. 12.

particular views you've developed over time allows you to easily dismiss any evidence or logic that doesn't correspond.

How many people in an organization will recognize when decision-making is driven by personal assumptions, opinions, and beliefs rather than data and logic, and how many will disagree with the opinions that form the basis of the decisions? Many, perhaps even most. What will this do to the broader perception of fairness in the overall decision-making process?

Let's recall from the early chapters the importance of the word "the" in the definition of value. No individual assumptions or opinions are relevant in determining either the expected future free cash flows or the OCC, the two essential elements of the value equation. When your assumptions about the future cash flows change, have the expected cash flows actually changed? Of course not – but it is at this point in the discussion that confusion often sets in. We are not making the claim that management has no impact on the future free cash flows. How managers run the organization, and therefore their behaviors and decisions, profoundly influences the expected cash flows. Their assumptions and opinions about the impact of these decisions, however, have no relevance.

Because each of us holds his or her own opinions in such high regard, we would like to believe that they should be the ones guiding our decision-making. As the biggest fans of our own opinions, we want them to do well and be important. In order to resolve whose opinions get used, someone must be handed control of the process. A hierarchy is needed.

So why do hierarchies get such a bad rap? What's wrong with them? More to the point, what's unfair about them, if anything? And what's so bad about organizing decision-making on the basis of a pecking order of opinion?

In any organization, the opinions of some people are implicitly placed above those of others, and therefore considered more valid. Your opinions will differ from those of every other person on the planet, at least in some respects. Does this mean they're better than everyone else's or worse? The answer is beside the point. The crucial matter as far as the blue line goes is that people will reject as unfair any system in which certain peoples' opinions are deemed superior and others second-rate. Indeed, any time an allocation of resources (jobs, promotions, bonuses) takes place on the basis of anything that isn't strictly data- or logic-driven – including assumptions, opinions, and beliefs – it will be perceived as unfair, and rightly so.

When you find yourself in a clear hierarchy and you're not at the top of it, how do you behave? For example, assume that the decision-maker one level above you has an opinion on an upcoming decision, and in your organization, opinions are not allowed to drive decisions but encouraged as important influencing factors. You have concrete evidence suggesting your boss's opinion is flawed. Do you present this evidence?

Let's consider the case in which you do raise these facts, confronting your boss in public or private and pointing out that her opinion is flawed because it just doesn't fit the data. What happens next? Perhaps your boss immediately recognizes your insightful and valuable contribution and dismisses her own faulty opinion, correcting the decision so that it conforms instead to the evidence and logic you so have diligently brought to her attention.

Alternatively, your boss points out that it is your data that is unfortunately flawed. *"It's from the past,"* she might say, *"and it's incomplete because you failed to gather evidence that would have supported my opinion."* On this basis, the boss, in addition, concludes that you are not a team player and that you are undermining her authority, thereby sowing distrust in the minds of your colleagues. Her decision sticks just as if you had never presented your case in the first place – and, oh, by the way, you've damaged your career in the process.

We suspect that the majority of readers will recognize the second scenario as the far more likely one. In most companies, speaking up or telling the truth, and providing evidence and logic to support it, is unlikely to either influence decisions or advance your career. The natural result is that very few people in such a system will tell the truth or offer any evidence that ever conflicts with the assumptions, opinions, and beliefs of those higher up in the organisation chart.

What about the implications for those at the top of the pyramid? At a certain point and a certain level, it will have been years since anyone inside the company has told them the truth. Put another way, any "evidence" they've been offered will have been information confirming their opinions. What do people at these levels come to believe? That their assumptions are superior, since years of "data" have confirmed it.

Taking these points together, we arrive at a disastrous conclusion. The least knowledgeable people are driving the allocation of energy inside the company, and eventually everyone will come to recognize this. Will such a process be perceived as fair? Of course not. No trust will be placed in the integrity of the decision-making process or the competence of the decision-makers. Data is relegated to worthlessness, and telling the truth represents the biggest danger. The blue line is obscured; value, ignored.

## Visions, Missions, and the Blue Line

There was a time when the "visionary" leader was all the rage. Such sages are still written about with great affection and admiration in the business press. Great vision, after all, leads to powerful mission statements, which allegedly give a company its direction. The mission statement also assumes the dual task of inspiring employees and lifting the organization out of the ordinary into the stratosphere.

We see it a little differently: if you have a vision, go see a doctor. Why are we down on visions and their accompanying missions? Because they run counter to the blue line. First, visions rarely speak for the entire organization. Because of this, they imply that someone is in control and that this person doesn't care what you say – if you don't agree with the vision, it isn't going to be fun to be you. Those who don't "feel" the vision inevitably end up waiting to be told what to do rather than taking initiative.

Second, leaders with a vision tend to wear blinkers in pursuit of it. They suffer from what psychologists call "confirmation bias," the tendency to see or hear only those things that support one's own opinions. Vision-driven leaders tend to pick up on all confirming signals and miss all information to the contrary, even if that information is coming in the form of clear signs from the market, customers, employees, or suppliers.

Third, and worse, the vision mindset violates one of the logical cornerstones of fair process: the willingness and ability to change. When other members of the organization realize that the boss is insistent on clinging to the vision because it's nailed to the lobby wall and hung in a nice frame, even though the signs are clear that the vision is outdated, outmoded, or just plain wrong, they aren't exactly going to fall in line. On the contrary – they will become demoralized and indifferent. The power resides at the top, and they have no hope of influencing change through data or logic, so they instead become resigned – the worst possible state for any organization's people.

Finally, visions and mission statements, in the end, rarely produce useful insights that might contribute to a practical strategy. When he gave his first press conference after taking the helm at IBM, Lou Gerstner famously said, *"The last thing IBM needs right now is a vision."*[16] Vision statements may work if they're designed to build a sense of common purpose. In such cases, they can help mobilize and motivate people, but when they cross the line from "why" to "how," a dichotomy we will explore in the next chapter, they are more likely to provoke resentment and discontent.

For the most part, it's easy to agree on an organization's "why." As an example, W. L. Gore & Associates proclaims its purpose as *"make money and have fun doing it."* But nobody can possibly know in advance exactly how that objective is to going to be achieved. Therefore, the organization must allow its members to explore various paths, through experimentation and learning, without their bosses telling them how to satisfy the vision.

The suggestion that the guy at the top knows how, and what, each of us needs to do in order to achieve the common purpose, is obviously absurd. That suggests a level of insight and knowledge no one can possess. The "how" implied in the vision will certainly be wrong in at least some respects. However, because the CEO has staked his or her credibility on it, and because it's become enshrined in a formal and highly visible document, members of the organization can be counted on to manipulate information flows to confirm the vision, even when the facts suggest otherwise.

If the Blue Line Imperative asks that we bypass vision-driven decision-making, what should the value-based leader do instead? Listen. Interpret signals and respond. Enable others to access the resources and take the actions needed to respond quickly and accurately to customers. Pursue

---

[16] Gerstner, op. cit., p. 68.

positive-NPV decisions, and encourage everyone else up and down the ladder to do the same.

Every company, of course, needs a reason for existing and an aspiration to work toward. There is nothing wrong with putting those things in writing. Some companies, like Google (*"Do no evil"*), have decidedly non-specific but highly resonant mission statements, underscored by a distinct set of guiding principles that highlight value for the customer, fairness to all, and trust. Google's philosophy begins with *"Focus on the user and all else will follow,"* while W. L. Gore & Associates's guiding principles include freedom, fairness, and commitment. These are value-oriented companies in action, where the Blue Line Imperative can be both heard and seen.

CHAPTER NINE

# TRUST AND HIERARCHY IN BLUE LINE MANAGEMENT

*"Humans are deeply social beings. Most people prefer to be in company most of the time. . . . As social beings, we want to trust each other."*[1]

– Richard Layard, economist

As we have said, value destruction occurs when a company fails to allo-cate energy in ways that increase happiness for its customers compared to alternative uses of that energy. The way in which the business is run creates the presence or absence of trust, a key factor in whether value can be achieved. As we saw in the previous chapter, when people in an organization lack trust in one another, value destruction is inevitable. As they observe value-destroying behavior with no penalties for those responsible, the result is a spiral of ever-increasing mistrust in one another and in the system, and ultimately, the company's demise.

By contrast, the Blue Line Imperative requires a culture conducive to value-oriented behavior on the part of every individual in it. If the organi-zational structure is not designed for value, it will be impossible for the people operating within it to conduct themselves in a manner oriented toward the blue line. It is therefore imperative that we give careful atten-tion to the structure of our businesses, otherwise, we don't offer the

---

[1]Richard Layard, *Happiness: Lessons From a New Science*, (London: Penguin Books, 2005 (2011, revised edition)), pp. 225–226.

platforms that allow people to create value. The absence of trust doesn't lead to value destruction; it *is* value destruction.

The role of trust in promoting civilization, and allowing humanity to develop and progress, has been well established. There is a fundamental link between having trust among individuals within an organization and that organization's ability to manage on the basis of value creation. Mutual trust across the organization allows everyone the freedom to pursue value. But this philosophy must be fully endorsed, thoroughly communicated, and completely shared.

## Why Trust Matters

In an interesting thought experiment, Jonathan Haidt, a specialist in moral psychology at the University of Virginia, asks what might have happened if a group of aliens had dropped ploughs, seeds, and pictorial instruction cards all over Africa one million years ago. Would our distant ancestors, having the right tools and guidelines, have become farmers? Haidt describes why he believes this almost certainly would not have happened:[2]

> "It takes far more than technology to farm; it also takes
> cooperation. Many people must work together with such high
> levels of trust that they can divide up the tasks and then toil
> for many months with no reward. When the harvest finally
> comes in, the farmers must be able to share it, store it, defend
> it, and make some of it last until it's time for next year's
> planting."

In short, it would take many millennia before human kinship provided our ancestors with the social organization that made agriculture possible.

And to go beyond simple agriculture, to build cities and nation states and the platforms that deliver the goods and services that make us happy,

---

[2]From a pre-publication version of Jonathan Haidt, *The Righteous Mind: Why Good People are Divided by Politics and Religion*, (New York: Pantheon Books, 2012).

we needed systems of organization that went beyond family and clan. We needed some form of glue to bind us together to do complex things. It wasn't physical or intellectual ability. It was trust.

It's easy to see at a basic level how important trust is to the average business relationship. In the trading pits of the Chicago Mercantile Exchange, for instance, or in just about any major commodities market in the world, if you develop a reputation for being less than entirely trustworthy, perhaps by having reneged on an earlier trade, other traders will ignore you. You can shout and scream your orders at the top of your lungs, but having shown even once that your word is not reliable, it's likely you won't be listened to, or responded to.[3]

You cannot do business without trust. Or perhaps we should generalize and say you cannot do business *efficiently* without trust. Go outside the commodity exchanges to a typical office in which a deal between parties is going down, and you quickly realize how much longer it takes when those involved have between them even a modicum of mistrust – and by the same token, how quick and effortless it can sometimes seem when the relationship is strong enough for the deal to be sealed with a handshake. As economist Paul Zak puts it, *"Trust is among the most powerful stimulants for investment and economic growth that economists have discovered."*[4]

A culture of trust is vital to the blue line organization, and fairness, as we have noted, is an absolute precondition. If your employees see that you operate in a fair and unbiased way, they can take a long view, trusting you to do the right thing by them and in return doing the right thing by the company. If they think you are behaving in a short-term, small-picture way intended to serve the agenda of certain individuals, or of a strategy they instinctively see as value-destroying, they will respond accordingly.

Robert Levering distinguishes between organizations that have "commodity" interactions and those that have "trust" interactions with their

---

[3]The one exception to this rule is if other traders sense that the guilty party has made a critical error. They will then pounce, and take his money in the process.
[4]P. Zak in M. Shermer, *The Mind of the Market*, (New York: Henry Holt and Company, 2008), p. 178.

employees.[5] He describes commodity interactions as "this for that." Employees provide a certain number of hours of their time, and in return the company gives them a biweekly paycheck. Trust interactions, on the other hand, are described as a "gift" economy, where trust and partnership, rather than direct payment, are the much healthier foundations for exchanges.

We agree with Levering's characterizations, though we portray them slightly differently. For our purposes, a "transactional environment" is one in which each request (for cooperation, help, or resources) is viewed as a standalone deal, and in which the person being asked seeks compensation before he is willing to undertake the requested action. In contrast, a "trust-based environment" is one where requests are honored and fulfilled in the knowledge that requests for help and cooperation will be reciprocated. We'll let you decide which of these is oriented toward the blue line and which toward the red.

Table 9.1: The differences between transactional and trust-based environments

| Transactional Environment | Trust-based Environment |
| --- | --- |
| One-for-one exchanges | Fair exchange works itself out over time |
| Relationship must be renewed after each exchange | Open-ended relationship |
| Avoidance of failure | Willingness to take risks |
| Short-term perspective | Long-term perspective |
| Absence of team mentality | Team-based perspective |
| Zero-sum game | Everyone wins |
| Indicator-based | Value-based |
| Risk of negative motivational spiral | Positive motivational spiral |
| Compensation is in the form of money only | Compensation is in the form of money, trust, recognition, and many other factors |
| Compensation plans commoditized | Rewards highly personalized, tailored to value contribution |

---

[5] Robert Levering, *A Great Place to Work*, (New York: Random House Inc., 1988).

A trust-based environment is, of course, harder to implement. It requires patience, faith, planning, and a highly specific organizational culture geared toward such a strategy. Levering cites the example of American managers trying to adopt Toyota's Quality Circle program, but failing because they approached it from a commodity perspective, that is, they were always focused on wanting to see the return on investment for each intervention. From a blue line perspective, it is the opposite view that would have been the productive one. The Quality Circle approach works only if there is an underlying commitment to the long term – to the fact that the organization will ultimately perform better in the future, even though stopping the production line or running an additional training workshop today might cause what feels like a temporary loss in value.

Such an approach also depends on a culture that considers all employees, even those at the bottom of the pyramid, as more than just cogs in the machine. The Blue Line Imperative says that for value to be created throughout the organization, it must first be transmitted through the people in the organization. Either everyone feels it, or no one does.

A broader way of saying this is that, in a true trust-based environment, in which the long-term pursuit of value holds sway, even when employees are asked to do things that may not appear right to them for the company, they do it anyway. In trying to explain this, Jean-François Zobrist, former CEO of French autoparts company FAVI, distinguishes between "how" companies and "why" companies. In the former, there are rules and procedures for everything, employees are encouraged not to think for themselves, and changing the way anything is done means writing a new rulebook. In other words, it's more about the rules than the customer. In the latter, the focus is on keeping the business's customers happy. If an employee asks why something is done the way it is, the only proper answer is that it makes the customers happier.

At Toyota, the fundamental value proposition is represented by a simple idea: value for the customer. That means *everyone* in Toyota is responsible for quality, no matter where they sit on the organization chart. All management systems and processes that make up the famous Toyota Production System (TPS), are oriented around enhancing value for the customer and nothing else competes with this theme.

In fact, when any system or process is examined within the Toyota walls, the first question asked is always the same: *"What does the customer want from it?"* The question shows Toyota's focus on value and also explains the messianic zeal with which it tries to purge waste from the system. Waste at the company is defined as any effort that doesn't add value – from, yes, the customer's perspective.[6] Examples include producing items for which there are no orders, workers standing around not doing much because of stockouts, taking unnecessary steps to process something, wasted motion by workers, and defects.

It all seems simple, doesn't it? Sure – but it isn't, because establishing and maintaining a blue line perspective is no easy task. Toyota certainly looks like one of Zobrist's "how" companies, but it works hard to maintain a culture in which each process is defined with the customer in mind, and it has over years developed a clearly defined manufacturing process that reinforces this value-oriented approach. To some, it may look like Taylorism – where everything is so minutely defined that the employee performing the function is simply an automaton easily replaceable by someone else with similar training. However, because of the customer focus and a culture that truly empowers every employee to try to do things better, processes are seen only as the best way the company knows how to do a certain thing today. Ideas for improvement and innovation – blue line staples – are encouraged by everyone, all the time. If an employee has an idea to improve a process, they can go ahead and see if their idea works. There is no mandate to stick with something just because of that terrible rationale, *"We've always done it this way."*

Does your company have a trust-based culture? Does it empower employees to do what is right for the customer, which will ultimately be right for the company and therefore create value? Or does it seem instead to operate on the basis of transactions, in which employees are asked to undertake value-destroying behaviors because there is no direct link to the customer? You know by now what happens in these types of environments. Employees do what they are told even if they don't like it or are suspicious of it, morale suffers, everyone goes through the motions to keep their jobs, they become less productive and less motivated, and the blue line nosedives. But in a culture based on trust and fairness,

---

[6]See J. Liker, *The Toyota Way*, Chapter 3, for an extensive discussion on how Toyota defines and manages waste. Jeffrey K. Liker, *The Toyota Way*. New York: McGraw Hill, 2004.

people feel the freedom to try and to fail, because their eyes are all on the same ultimate goal: value.

Timpson is a UK retail chain that specializes in shoe and watch repair, key cutting, and locksmith services. When Timpson buys another store, one of its first acts is to remove the electronic point-of-sale machines, replacing them with old-fashioned cash registers.

Why does Timpson do this? For one reason only: to signal to the shop employees that they are in charge and trusted to do what is right for the company.[7] If this means discounting certain products to move them out the door quicker, they can do it without getting anyone's permission. Such value decisions reside in their hands, and they don't need to refer to a manual or guidebook to figure out what they are or aren't allowed to do.

Not that Timpson doesn't have a training manual; it most certainly does distribute one to all new employees. To further strengthen its culture of trust and autonomy, however, the company includes in this manual a list of 20 ways to "rip off" the company as a means of letting its employees know where it is most vulnerable, yet it trusts them anyway. (A similar story has become part of corporate lore at Hewlett-Packard. One day, when Bill Hewlett found the door to the supplies room locked, he cut the lock with bolt cutters and left a note saying, *"Don't ever lock this door again."* The message was simple and clear: if we don't show our people we trust them, what do we have?[8]) As Tim Harford explains, *"Many people respond to being trusted by becoming more trustworthy."*[9] Timpson builds on this trust by allowing and encouraging its employees to recruit their own new colleagues, in what they call a "Refer a Friend" scheme. In effect, employees are trusted to select teammates who reflect their own value orientation, ensuring that the right kind of people are continually hired and thus the blue line culture continually reinforced.

A similar philosophy reigns at the US supermarket chain Whole Foods. There, the basic organizational unit is not the store, as with Timpson, but the team. Each Whole Foods store has, on average, eight

---

[7] Tim Harford, *Adapt: Why Success Always Starts with Failure*, (New York: Farrar Straus and Giroux, 2011), p. 227.
[8] Sally Bibb, *The Stone Age Company*, (London: Marshall Cavendish, 2005), p. 109.
[9] Harford, p. 228.

groups, and each of these has responsibility for all key operating decisions, including pricing, ordering, in-store promotions, and most important, staffing.[10] New hires are assigned to teams, given a four-week trial,
and then voted on by the team members. A two-thirds majority is needed
for a full-time spot to be earned.[11]

In addition, team leaders, in consultation with the store manager, are
free to stock whatever products they feel will appeal to local customers.
Needless to say, this model represents a radical departure from the usual
retail practice, in which buying decisions are made at the country level.
Not only does this approach make Whole Foods highly sensitive to the
wants of its customers, but it reinforces the company's trust-based culture,
and in turn its shared value orientation.

Besides granting this seldom-seen power to its internal teams, Whole
Foods builds trust in other ways. For example, it allows every employee
access to every other employee's compensation data. Its operating and
financial data (including daily store sales, team sales, store profits, and
so on) are available to any associate wishing to see them. It is a
"no-secrets" management philosophy that not only facilitates effective
decision-making at the team level but also serves a larger blue line
purpose. Gary Hamel says ". . . *open books are the only way to build a
company that is bound by trust. It's standard practice at many companies
to conceal information as a way of controlling employees – a formula
that's toxic to trust. By contrast, the top team at Whole Foods believes you
can't have secrets and have a high-trust organization.*"[12]

In their book *Toyota Culture*, Jeffrey Liker and Michael Hoseus discuss
Toyota's "trust bank." Describing the concept, a senior Toyota executive
says, *"If you do not give them your best, there is no reason for them to
give their best to you. You have to make deposits before you can make
withdrawals.*"[13] At W. L. Gore & Associates, a similar philosophy exists,

---

[10]Teams are organized around broad product groupings such as seafood or
produce, or critical store functions, such as checkout.

[11]Gary    Hamel, *The Future of Management*, (Boston: Harvard Business School
Press, 2007), p. 72.

[12]Ibid., p. 74–5.

[13]J.K. Liker and M. Hoseus, *Toyota Culture: the Heart and Soul of the Toyota Way*,
(New York: McGraw-Hill, 2008), pp. 366–367.

referred to as a "credibility bucket." According to the metaphor, each time an associate keeps a commitment, a drop of water is added to the bucket and, over time, trust accumulates. It takes years to fill the bucket, but only one serious breach of trust to poke a hole in the side, causing the entire thing to empty.

Trust affects every facet of a company's value chain, not just its employees. In the late 1980s, influenced by Toyota's success in managing its supply chain, General Motors began making a series of moves to establish strong, long-term relations with a select group of suppliers. In the interest of cooperation and engagement, the company persuaded these suppliers to share cost information they would normally have kept confidential. Unfortunately, when GM then got into financial trouble in the early 1990s, they used the information they had been given to force large price cuts from those same suppliers. Not surprisingly, as economist John Roberts notes, *"The breach of trust soured relations with the suppliers for years afterward."*[14]

If you want to observe first-hand what happens when trust disintegrates entirely, visit crisis-ridden Greece. That's what Michael Lewis did recently, and what he recounts is a numbing tale of bad intentions – just about everywhere he looked. The Greeks are a famously warm, funny, and smart people, in particular with foreigners. But try to get one Greek to compliment another,[15]

> "No success of any kind is regarded without suspicion. Everyone is pretty sure everyone is cheating on his taxes, or bribing politicians, or taking bribes . . . And this total absence of faith in one another is self-reinforcing. The epidemic of lying and cheating and stealing makes any sort of civic life impossible . . . The structure of the Greek economy is collectivist, but the country, in spirit, is the opposite of a collective."

---

[14]John Roberts, *The Modern Firm: Organization Design for Performance and Growth*, (Oxford: Oxford University Press, 2004), p. 207. In defense of GM, Roberts reports that some GM executives credited the money squeezed from the suppliers with saving the company from bankruptcy.

[15]Michael Lewis, *Boomerang: Travels in the New Third World*, (New York: W.W. Norton, 2011), p. 55.

In other words, no one trusts anyone, with the exception perhaps of immediate relatives and close friends. Everyone takes what they can, when they can, because they can, without regard to the economic or social consequences. Quite simply, as Lewis says, it's *"every man for himself."*[16] Greece in its current state is truly an example, writ large, of what happens to any organization when trust is absent.

You may recall our earlier discussion on lying, and why people feel bad when they do it. The Greeks are no different from the rest of us insofar as they don't enjoy the act of telling a lie. But at the moment they reside in a system that compels them to lie – and cheat, and steal. Let's be clear on the point. This is not a comment about the quality of the people in an organization – in this case, the Greeks. It is about the nature of a system – in this case, Greece. If everyone in the system observes everyone else lying, cheating, stealing, and profiting from it, implicit unfairness is revealed, producing a vicious cycle, because if my neighbor is lying, cheating, and stealing, then to avoid being duped or taken advantage of, I have little choice but to do the same.

As we write these words, the European Union, and more specifically the Euro Zone, is struggling mightily with the consequences of a bankrupt Greece. Everyone in Europe understands that a disproportionate part of the burden for bailing out the Greeks (not to mention the Irish, Portuguese, Spanish, and Italians) will fall to the Germans. The German people are understandably resentful of having to bear this burden, but they derive a certain smug satisfaction from telling anyone who will listen that the Greeks should be more like them.

At the same time, the Germans would be well served to reflect on their own recent history. Not long ago the eastern half of the country found itself in a system where lying, cheating, stealing, and even worse were endemic. In less than a generation, millions of Germans had adopted the same types of behaviors reflected by the Greeks today. Was there something fundamentally wrong with the people in East Germany *per se*? Of course not – no more than there was with the people in today's Greece, or in any red line organization.

It's much simpler than that. Any group, when they find themselves in a situation where lying, cheating, and stealing are the primary route

---

[16] Ibid.

to success, will first recognize the implicit unfairness of the system, and will then go against their natural inclinations and begin to lie, cheat, and steal themselves, because no other alternatives to survival exist. As one observer notes, *"No one Greek broke the state. An individual person or family simply found themselves within a system that encouraged cheating and exploiting from the state and discouraged honesty or self-sacrifice."*[17] In such a system, our most basic reflexes take over. We do what we have to in order to survive.

## Fairness, Trust, and Information Flow

Even in blue line organizations where fairness and trust are upheld, creating unimpeded information flows is a never-ending challenge. You've no doubt heard of the game "Broken Telephone," or, if you're British, "Chinese Whispers" (or maybe you call it "Grapevine," "Whisper Down the Line," etc.). This game offers a great example of how tricky it can be to get information to where it is needed without losing any of the details or original meaning. In red line organizations, broken telephone runs rampant because people purposely hide information from one another. In a blue line environment, that isn't the intention, but it is nonetheless difficult to manage.

A probably apocryphal story tells of a message sent by a soldier in World War I, *"Send reinforcements, we're going to advance,"* which emerged as, *"Send three and fourpence, we're going to a dance."* But the point is relevant. It's easy to see how this sort of thing might occur in any company today. Think of some junior staff in a bank. They've got their ears to the ground and are getting worried about what customers are saying – objecting to the large fees the bank charges to move money around, pointing out that there are Internet or Smartphone solutions that are much cheaper, and so on, and perhaps they'll leave.

What the junior staff are hearing sounds like, *"Money wants to be free,"* but the bank is a hierarchical business, and the message has to work its way up the chain to the decision-makers. By the time it gets

---

[17] Max Fisher at http://www.theatlantic.com/international/archive/2011/11/the-only-leader-who-understood-greeces-real-problem-is-resigning/248018/

there, the message has turned into *"Mommy wants a Wii,"* and the leaders shake their heads. An organization like this cannot learn because it does not receive the signals clearly from its customers, nor can it respond quickly to changing customer moods because it takes too long to understand what they want.

The significance of properly identifying and interpreting the flow of signals emanating from customers cannot be overstated. To get the right stuff to the right people requires that your company is not only aware of its customers' desires but can read them properly and deliver on them in an effective manner. Remember, you can only be said to be creating value if your customers feel you are, so the only thing that really matters is that you make sure you know what those customers want and what value will look like to them, meaning of course, what will make them happy.

For this, you need to make sure that information – for example, customers don't like blue widgets – gets quickly and accurately to where it's needed and can be acted on. Today's business environment requires much greater responsiveness and speed than ever before. Organizations that pick up signals quickly and clearly from the market and their customers have a distinct competitive advantage over those who hear less and react more slowly.

There are two main hurdles to getting information around an organization to the right places. First, as mentioned above, messages tend to get distorted. Second, information of strategic value often gets protected. Either circumstance means the company is less able to respond to signals from its customers, and less able to create value.

## Getting the Right Information to the Right People, Fast: The Case of Alcoa

Some companies, in their efforts to espouse the Blue Line Imperative, get it right. A sterling example is Alcoa, a supplier of aluminum and other materials to businesses around the globe, which created an organization-wide information sharing system that shifted its safety record from "good" to "amazing."[18]

---

[18] For a comprehensive description and analysis of Alcoa's information system, see Steven J. Spear, *Chasing the Rabbit*, (New York: McGraw-Hill, 2009), pp. 87–104.

When Paul O'Neill arrived as CEO of the company in 1987, he announced that safety would be its new number-one priority, and that the organizational goal would be zero injuries. This statement had a number of implications. O'Neill's view was that any accident indicated that something in the company wasn't being done perfectly, that is, some piece of knowledge was missing. If knowledge was missing, then the company wasn't working as efficiently as it might be, or quality might be lacking somewhere. A safety focus also reinforced the basic premise that Alcoa didn't want to hurt people, that it was moral.

To capture knowledge quickly and in a reliable, trustworthy way, O'Neill insisted on receiving a full report on all accidents within 24 hours. This not only served to increase speed of response to accidents, but it also meant that as much relevant information about how and why the accident had happened could be captured before it was forgotten.

In addition, and perhaps more important, anyone who might typically be tempted to use or withhold information for personal advantage would now find such behavior to be, if not wrong, at least vastly more difficult given the new requirement to move information more quickly. The 24-hour window would mean that only the facts could be passed on, with little time for manipulation or hoarding of information. Apart from the 24-hour report, O'Neill also asked that, within two days, he receive a second one detailing what was being done to make sure the specific type of accident wouldn't happen again. This information was shared across the company. Others encountering similar problems could now learn about steps they could take to solve them. Overall, the new safety focus allowed the company to learn about its processes in enormous depth, creating a culture in which it was expected that processes and issues were to be examined in detail, for one purpose: helping the company understand where it could do things better and therefore create more value.

What we're really talking about in the case of Alcoa's blue line culture is its focus on continuous learning. In a value-oriented culture, fairness and trust reign supreme, information is exchanged freely, and everybody learns. Because of this fairness and trust, people feel at liberty to share their mistakes as well as their successes. As we've said, only in a red line organization is the focus squarely on successes. In a blue line company, by contrast, employees feel safe enough to share when things don't work. When there is an accident, or performance is deemed insufficient in some way, for example when a department falls short of targets, value-creating companies don't look to apportion blame. Instead, they seek explanations by asking the right questions. What have we learned from this? Were our expectations unreasonable? Was there support that we should have provided – IT solutions, additional staffing, training – to help the manager perform better? And if there is blame to be assigned, no automatic assumption is made that the manager in question is deficient in some way. The more important question is asked: what did we as a company not do, what did we not know, and what support should we have given that we didn't?

## Hierarchies and Information Flows

A vital element in trust and information flow within a company is its structure. Traditional, steeply hierarchical organizations suffer most from distorted information and bad news being swept under the carpet. They are slow to respond and are less likely to know what their customers really want.

Imagine that the top executives at a big automaker have formed the hypothesis that family sedans are more profitable for the company than small, energy-efficient cars. Suppose the opposite is in fact true: customers actually prefer small, energy-efficient cars and are willing to pay a premium to get them. To correspond with the company's efforts to generate high profits, and the mistaken hypothesis that it can do so by selling large sedans, it motivates its salespeople to sell more family sedans, perhaps by setting sales targets and tying bonuses to delivery on these targets.

What happens then? The sales channel moves aggressively to hit the targets, probably by offering incentives to customers like price reductions, free options, or low-cost financing. These incentives, of course, have a negative impact on profitability. As the salesforce starts hitting the targets, the hypothesis that customers want family sedans is confirmed, and the decision not to sell small, energy-efficient cars validated. Customers may be screaming about their preference for smaller cars, but the salespeople don't listen because that isn't what they're paid to be doing.

The domino effect continues. Senior managers will see that they can sell the sedans (after all, the sales force is hitting its targets), but they will also notice that the company seems to have moved from a profit-generating position to a loss-making one. In no way will it occur to management that the root of the problem is their original hypothesis about what their customers want. The difficulty, as they see it, is that costs are too high – so they embark on a major cost-cutting initiative. However, because they are selling the wrong cars, it is virtually impossible to cut costs sufficiently to return the company to profitability. Moreover, they never receive signals from the sales force that customers actually want small, energy-efficient cars. Any signals coming from customers are obliterated in an avalanche of KPIs and misdirected incentives pushed from the top down.

In a blue line culture, the emphasis on fairness and trust lets people know that there need not be excessive layers of management. Employees understand that they don't need to be constantly looking over their shoulders for someone making sure they're doing the right thing. They will do it anyway.

We were recently reminded of an unfortunate example by a former employee of SAirGroup (the holding company that owned Swissair). Until the late 1990s, Swissair maintained a well-earned reputation as one of the world's best airlines. However, in 1998, the company embarked on an all-too-familiar strategy that focused on growth and pinned mainly to the acquisition of minority stakes in European airlines (to avoid loss of EU status and associated landing rights), financed with debt. These acquisitions proved profoundly value-destroying and would ultimately lead to the company's demise by 2001.

In the year leading up to the collapse, certain people inside the airline noticed the sharply deteriorating cash position. Investors noticed too, and the company's share price went into freefall. Nevertheless, senior executives remained stubbornly committed to growth while appearing mostly oblivious to the company's impending bankruptcy. As an example, at the beginning of 2001 (the company would be dead within 10 months), the CEO announced that the company had earned SFr200 million in profit in 2000. The actual number, released only weeks later, turned out to be a SFr2.9 billion loss.

How could the CEO have been so confused? A participant in one of our executive programs, and a former employee in the SAirGroup's treasury function, recounted the months leading up to the collapse. According to her forecasts of future cash generation and the cash needs of the company, it became apparent that the company was heading for disaster. She had gone to her boss at SAirGroup and attempted to show him the relevant numbers, but was dismissed with little consideration and reminded, *"We are Swissair; we aren't going to default. And I don't need to see your numbers to know this."*

Following repeated futile attempts to get through to the boss, this employee decided to go over his head and arranged an appointment with the Group CFO who explained to her, *"We aren't going to default, we're Swissair, and if we were going to default, your boss would have already told me. I don't need to see your numbers to know this, so you need to be more careful in your calculations."* He added a further comment: *"By the way, it is unacceptable to violate the chain of command as you have done. If you wish to say something to me, do so through your boss."*

With only one choice left, she tried to get an audience with the CEO to present her calculations. By now, default was starting to look unavoidable, but she felt she had to try to make someone aware of what she'd found. She managed to book time with the CEO and tried her best to persuade him to reverse course in the hopes that disaster might still be averted. His response? *"You must be confused. We are Swissair, we aren't going to default. And if we were going to default, the CFO would have told me. No, I don't need to see your numbers, I'm quite sure you've made mistakes."* He too, felt it necessary to add a bit of editorial: *"By the way, you have been quite bold to approach me this way and not go through the appropriate channels. In future, if you wish to send me a message, please do so through your boss and his boss, so that I don't waste my time on issues that they have already determined are not serious enough to bring to my attention."*

The CEO lost his job a few weeks later when the financial condition of the company and the impending disaster became known to the public.

The Swissair example offers just one of many poignant cautionary tales highlighting how damaging a strict hierarchy can be in stifling information flows, compromising trust, destroying any possibility for learning, and obscuring the blue line all but completely. As Swissair's failure was becoming unavoidably clear, SAirGroup Chairman Eric Honegger told *Neue Zürcher Zeitung, "In retrospect, we determined that problems were much bigger than people believed at the time, that much more money had to be invested, that the leadership problems were bigger, and that our influence in the foreign participations via these complex constructions could be realized insufficiently."*[19]

The key word in Honegger's quote, and the one that screams red line, is of course, "believed." It was evident from the former employee's story that the executives were willingly and blissfully ignorant of the data and of the true state of Swissair due to their misguided belief that the company was invincible. It simply couldn't go bust. In their minds, therefore, the data was irrelevant no matter what story it might purport to tell. They knew the truth because of the force with which they believed in it. In a hierarchy such as that which sent Swissair crumbling from the top down, people at the top have an unfortunate power to squash data using nothing more than belief, and there is little to nothing others can do to fight back. The often tragic result is that there is no reason to gather data in the first place, any possibility of learning disappears, and the only way to go is down.

But let's not be so pessimistic. We can also point to numerous cases of companies that get it right. As we've mentioned, W. L. Gore & Associates is among our favorite examples of companies whose commitment to fairness and trust is in large part why they are able to constantly maintain a robust blue line and create ongoing value and sustainable competitive advantage. At W. L. Gore & Associates, learning is king, and the only goal that matters is the collective drive toward value creation.

For starters, there are no managers or employees – everyone at the firm is an associate. When we met Frederic Amariutei, Directeur Général (DG) for W. L. Gore & Associates in France, he told us he wasn't the boss. He just had the title because when he met with other leaders in

---

[19] Chairman Honegger would later say that *"Despite all the problems with the (foreign) participations, which are big but localized, there is (still) Swissair. One should not believe that everything has collapsed."* This statement indicates an emotional connection to the brand "Swissair," which prevented him, as it had other senior executives, from seeing the data clearly. All quotes are from Winfried Ruigrok, "A Tale of Strategic and Governance Errors: The Failings Which Caused the Demise of Swissair were Aggravated by the Convergence of Several Industry Developments," *European Business Forum*, Spring 2004.

the French division, they agreed that he would make a good DG. *"I only assume this title for legal purposes,"* he said. *"Internally, I am an associate, just like the others. I know that I am not the 'boss' of anyone. I am a member of a team of people who manage the French activities of the company. We, together as a Leadership Team, are the Directeur Général, and I merely have my name there because under French law someone's name has to be there."* [20]

As an extension of its non-hierarchy, everything at Gore is done by teams, and the formation of these teams follows no pre-set pattern. The teams form as they are needed, to manage ongoing operations or particular projects, and then dissolve when they are no longer needed or when the team realizes that what seemed like a good idea is never going to come to fruition.

In the latter case, Gore does something surprising, as we learned from a participant in one of our top-level programs who had recently visited the company's global headquarters in Delaware. When he arrived, he found a party going on. He asked his hosts what they were celebrating – a new client acquisition, or perhaps the launch of a new product? No, he was told, they were celebrating the closing down of a failed project. Our visitor was surprised, but further explanation revealed that, although the idea hadn't worked out, it had seemed like a good, value-creating project in the beginning. For reasons nobody could foresee, it hadn't worked out, but they were celebrating the fact that someone had generated the idea and that other associates had agreed and worked on it, and that everybody had tried.

They were reinforcing the notion that, at Gore, it is okay to fail, as long as the facts suggested that the idea was value-creating and something learned from it. This virtually defines a blue line culture. When later facts revealed that the idea was not working as expected it was stopped. A more typical corporate response might have been something harsher, more accusatory, and demoralizing. The people at Gore recognize that not everything succeeds, and instead they emphasize the principle that fear of failure shouldn't stop anyone from trying, because if it does, nothing ever gets done, and no value ever gets created.

---

[20] K. Kaiser and S.D. Young, *Hierarchy and Compensation at W. L. Gore & Associates & Associates*, INSEAD case, 2009.

Bill Gore, the company's founder, worked as a research chemist at DuPont for nearly two decades and brought the ethos he had learned into his eponymous company. Among the lessons he had derived at DuPont was that small, non-hierarchical teams were highly effective at getting things done, largely because team members, with their detailed understanding of the relative strengths of the other members, knew who best to ask for help or ideas. He also observed time and again that there was little or no opportunity for free riding in small groups; everyone had to contribute to the collective effort or they'd be asked to leave. When starting his own company, Gore spread the idea of small teams across the organization, using team members not only to deliver projects but also to rank contributions and determine compensation.

The flat structure of Gore, with its highly mobile employees and fluid team structure, ensures that lines of communication are direct and unhindered by procedure or rank. Leadership roles are fluid; an associate might lead one project but be a team member on another. As Gore people like to say, *"If you call a meeting and people come, you are a leader."*

But without managers or bosses, who are the decision-makers? Gore has developed the "waterline" concept to help associates determine whether they need to seek advice from others before making a decision. This determination is based on what would happen if the decision in question went wrong. They are to imagine that the worst possible consequence of the decision would be to punch a hole in the side of a boat. Where the hole is relative to the imaginary waterline is the critical thing. If it is below the waterline, the boat will sink straight away. Even if it is above the waterline, there could be problems should the weather become stormy.

New associates at Gore have a very high waterline. They need to consult on most of the decisions they make. Associates who have been with the company a long time and who have important leadership responsibilities have a much lower waterline – they have more freedom to make decisions without consulting their colleagues. But everyone leverages everyone else, no one makes decisions in a vacuum, and the ultimate criterion is value.

At Red Gate Software, a UK company that regularly features in *The Sunday Times'* Best Companies to Work For list, the blue line philosophy of learning and experimentation is enshrined in the organizational handbook, *The Book of Red Gate*. This approach stands out in high relief in statements like *"Visible misteaks [sic] are a sign that we are a healthy organization. What we do is very difficult, the current situation is hard to understand and the future is uncertain. Mistakes are an inevitable consequence of attempting to get the right stuff done. Unless we can make mistakes visible both individually and collectively we will be doomed to mediocrity."*[21]

Red Gate's message to its employees is crystal clear: making mistakes, learning from them, and having another go fit a value-orientation paradigm. Researchers know they may well not get things right the first time, but this is the very nature of experimentation and, consequently, of innovation. When something doesn't work, people don't raise eyebrows or point fingers. More important yet to the blue line, they don't pretend the thing worked when it didn't.

## Trust Leads to Quicker Decision-Making and Happier Customers

A Middle-Eastern company that builds and operates oil pipelines was looking for a product to help ensure the structural integrity of the pipelines they manage and provide warnings of leaks. The company approached a number of well-known, highly reputable oil services firms, but they were also made aware of a competing service offered by Gore, a passive diffusion sampler called the GORE-SORBER® Module.

As discussions with the potential service providers began, the Gore culture quickly became apparent to the prospective client and became a key factor in the ultimate purchase decision. Compared to the other potential suppliers, the contact person on the project described the Gore salesman as *"extremely accessible, extremely customer focused, and extremely responsive."* Even more impressive was the speed and accuracy the salesman displayed in response to technical, legal, or contractual questions.

When other potential providers were asked such questions, their typical response would be, *"Let me follow up on that with the relevant people inside the company and I'll get back to you asap."* The responses

---

[21] Red Gate, *The Book of Red Gate*, Red Gate Software Ltd, (UK: Cambridge, 2010), pp. 62–63.

tended to be professional, fairly quick, and usually delivered within a few days. When asked similar questions, the Gore contact would typically respond instantly with the answer. When asked if there were other people in the organization who needed to confirm or validate the answer, the Gore person would simply say, *"No, this is the answer, and I can confirm it."* The customer, never having encountered such a response before, found it hard to believe at first. But after several weeks of correspondence, he became convinced that indeed, the answer given by the Gore representative was accurate, and that his nearly instantaneous responses could be counted on as accurate and reliable.

When the decision was taken to proceed to the next phase with Gore, the client contact asked the Gore contact to bring his boss to meet the client contact's boss in their office in Abu Dhabi. When the Gore representative explained that there is no "boss," the client responded, *"I know, I know, you explained that you don't have a hierarchy, but please just bring your boss to meet my boss because that's the way it's done here."* It took the Gore representative several attempts before the client agreed to accept that he would be the one to meet the client's boss. Again, during these and future meetings, there was only the one Gore contact, but at each meeting this person was able to answer all contractual questions immediately. Gore got the contract and, not surprisingly, both sides have been very pleased with the results.

## The Future of Hierarchy

When we discuss W. L. Gore & Associates with participants in our executive programs, a typical response goes something like this: *"It's interesting, but I just don't see how that sort of thing can work in my own company."*

Before dismissing Gore and other flat-structure companies out of hand as interesting case studies but "not for us," ask yourself a question: *"Am I more or less empowered in my work than my father or grandfather was in his job 50 years ago? And 50 years from now, will my grandchildren be more or less empowered than I am?"* For the overwhelming majority of us, the trend is clear. Businesses have been de-layering – reducing the number of rungs between the CEO and the lowest-level workers – and empowering employees in ways that were hard to imagine as recently as our grandparents' generation.

This trend manifests itself in many ways, from the networking phenomenon common in the technology sector, where even the most junior staff can talk face-to-face with anyone else in the organization without having to go through multiple channels, to the rapid growth in the outsourcing of functions previously done in-house (and often to former company employees). Some of these forces at play are technological, such as web-based platforms that make it easier for people to work from home for a potentially limitless array of clients. Others are sociological or psychological, in a very real way reflecting our long-standing desire to be in charge of our own destinies.

It isn't hard to find companies that still follow a traditional form of top-down management. But businesses of the future will have to respond to these realities; those that fail to respond quickly enough will be killed off. We aren't saying that hierarchies are always bad, or that all forms of hierarchy are inconsistent with the demands of a value-driven culture. Certain types of hierarchical structures can help clarify responsibilities, as long as decision-making is based on data and logic, and not on assumptions and opinions. It is when organizations fail to differentiate between hierarchical approaches to organizational structure, and hierarchical approaches to relationships, that problems arise. The former refers to decision-making authority and accountability, the latter to where "the boss gets to dominate all conversations."[22] In explaining this crucial distinction, UCLA professor Samuel Culbert writes,

> ". . . while hierarchy is essential for clarifying jurisdictions and responsibilities, it is entirely dysfunctional in a relationship and the development of that all-too-crucial trust. It makes healthy, give-and-take conversations all but impossible between bosses and the people who report to them. The reason is obvious: How can you have honest give-and-take when one person has a gun to the other's head, when the structure doesn't require, or even encourage, one party to listen to the other? When the structure, in fact, actively discourages listening?"[23]

---

[22] Samuel A. Culbert, *Beyond Bullsh\*t: Straight-Talk at Work*, (Stanford: Stanford Business Books, 2008), pp. 74–75.

[23] Samuel A. Culbert, *Get Rid of the Performance Review!*, (New York: Business Plus, 2010), pp. 7–8.

In short, when hierarchy is allowed to govern personal relationships, accountability becomes one-sided. Only bosses get to hold subordinates accountable, and they are also able to punish at will. Culbert's critique calls into question the all-too-common practice of dismissing subordinates who fail to hit the numbers while allowing the bosses who are responsible for coaching, guiding, and overseeing them to distance themselves from these "train wrecks." Such unhealthy relationships are reinforced by the use (and misuse) of performance reviews. While reviews are supposed to teach subordinates about what they should be doing better, Culbert claims that *"The real purpose is intimidation aimed at preserving the boss's authority and domination."*[24]

As a healthier alternative, he proposes two-sided accountability – relationships in which subordinates stand accountable for their performance and bosses stand accountable for giving the guidance and creating the conditions that allow them to succeed. *"In this view, the bosses' number-one duty as leaders is choosing subordinates who have the capacity to contribute to the organization and providing the tools and conditions for their success."*[25]

An important implication of this approach is that only by insisting on a "straight-talk relationship," with accountability on both sides, is it possible for an organization to become truly learning-based and therefore satisfy a vital element of the Blue Line Imperative. The consequence of failing to achieve the desired results, as Culbert points out, *"should be personal development, new perspective, improved judgment, skill enhancement, and general all-around learning."*[26] Subordinates learn what they need to do to improve performance, and bosses learn what type of support and guidance the subordinate needed but did not receive. When the subordinate fails, the boss does too, and both gain the opportunity for mutual learning and progress.

An interesting twist on two-sided accountability can be seen in what McKinsey & Company call the "obligation to dissent." The firm has a well-delineated hierarchy, from recent graduates to managing directors,

---

[24] Ibid, p. 3.
[25] Samuel A. Culbert, 2008, op cit. pp. 75–76.
[26] Ibid, pp. 76–77.

and everyone knows their place in it quite clearly. However, the firm's culture prevents the red line from taking over by preventing the decision-making hierarchy from getting in the way of effective problem-solving for clients. New hires are given countless inspirational examples of low-ranking consultants who dissented from the proposed recommendations for a particular client and were recognized and lauded for having spoken up. It is emphasized that, far from damaging your career, if you fail to exercise your obligation to dissent, your future at McKinsey is likely to be unhappy and short.

It is equally emphasized at the firm that when you exercise this obligation, you do so with sufficient facts and logic to explain why you dissent from your higher-ups, as opposed to simply saying you disagree because you think that's what you're supposed to do in order to be noticed. To ensure that the role hierarchy does not stifle the personal contributions of the firm's junior members, the obligation to dissent is continuously reinforced in training and intra-firm communications.

Pixar Animation Studios, the producer of popular animated films such as the *Toy Story* series (and part of Walt Disney Company), has a similar operating philosophy to that of McKinsey. Although dissent is not defined as an obligation, Pixar's top managers have, as one observer puts it, *"a relentless desire to challenge and learn and they ensure that this trickles [down] throughout the rest of the company."*[27] For this reason, everyone in the company is encouraged to disagree or express their dissent with any aspect of the business, including its creative work, without threat of penalty. *"And,"* adds one executive, *"because there's [no] penalty, people are more likely to say what they think."*[28] Trust is the rule, fairness the law, and value the constant and unwavering goal.

---

[27] Peter Sims, *Little Bets: How Breakthrough Ideas Emerge from Small Discoveries*, (London: Business Books, 2011), p. 41.
[28] Ibid.

# CHAPTER TEN

# VALUE AND DECISION-MAKING

*"Society functions in a way much more interesting than that multiple-choice pattern we have been rewarded for succeeding at in school. Success in life comes not from the ability to choose between the four presented answers, but from the rather more difficult and painfully acquired ability to formulate the questions."*[1]

– David Mamet, playwright

Without a value-oriented culture in place, business decisions cannot be value driven. At the same time, having the right culture in place does not mean decisions *are* value driven. While the three pillars of the Blue Line Imperative – fairness, trust, learning – are prerequisites to survival and success over time, they do not tell the entire tale.

Another way of saying this is that merely wanting to create value, or even creating the right conditions for it to flourish, is not the same as having a process that enables it. You can build the perfect environment for a garden, but if you don't know how to plant, your efforts won't amount to much. The first step is ensuring that a value-based culture is in place, built around fairness, trust, and continuous learning. The next step is to establish processes by which value-based decisions are made.

---

[1]David Mamet, *The Secret Knowledge: On the Dismantling of American Culture*, (Sentinel: Kindle Edition, Kindle Locations, 2011), pp. 452–454.

## Value and Physics

Earlier we described price as a point-in-time observation and value as the underlying, though unobservable, truth. As you manage your business, you make observations on a number of elements – sales figures, production quotas, supplier costs, employee turnover, and so on – and each of them might indicate something other than what you think they do.

Employee turnover numbers, for instance, might tell you about conditions in a plant, HR policy, salaries, or any number of other factors that contribute to high or low retention rates. If turnover numbers increase, then how do we see the underlying truth? How do we observe the unobservable and formulate the proper, value-creating response?

Some of the answer is available through the lessons of physics and a couple of its key figures, German theoretical physicist Werner Heisenberg, famous for his uncertainty principle, and Niels Bohr.

Both sought to get to the bottom of modern, quantum physics. Bohr was highly conceptual, visualizing the inherent structure of the atom and from there developing a deep understanding of the workings of nature. While Bohr visualized orbiting electrons around an atomic nucleus, Heisenberg decided that the idea of orbiting electrons was purely theoretical, since one could point to no observational evidence that they were doing so. On this basis he abandoned any attempt to visualize what was happening at the atomic level and instead focused only on what could be measured in the laboratory. Using this approach, Heisenberg developed his remarkable insights into nature's mystical behavior at its smallest levels.

Despite their contrasting approach, both of these towering scientists were able to accept that observations are not reality, but instead the crystallization of underlying forces. They understood also that our ability to understand, explain, and manage the underlying forces is strictly limited to what we are able to observe.

As we have said, the workings of business present us with a very similar problem. No matter what industry you're in, the indicators are the only observables. Generally speaking, rising profits are a fair indicator that value is being created. The same can be said for market share, customer satisfaction, employee satisfaction, return on invested capital, and all similar indicators. But we must always keep in mind that it is *what they are indicating* that matters, not the observations themselves.

Saying it in Heisenberg's language, limiting ourselves to understanding business from the perspective only of what is observable is not equivalent to assuming that what is observed is what we truly care about. Returning to our earlier discussion, what we care about is an understanding of the probabilistic nature of the underlying forces acting on the observable stuff, because only when we understand these forces can we begin to anticipate and manage the impact of a change in the underlying environment, whether atomic or commercial.

Our pursuit of an understanding of what we can do to impact value must therefore begin with the indicators, because they are our window into the true drivers of value – the ones we can't see plainly but which we must get at somehow. We can then turn to physics to help us ask the most important question: if we change certain inputs, how do the indicators change? As long as the indicators are not in some way biased, the answers to this question will start us on the path of understanding.

Some of you will no doubt recognize the similarity of this concept to that of Six Sigma, which asks us to assume a "black box" into which we feed inputs (behaviors and actions) and out of which we get outcomes (indicators), and whose ultimate goal is to understand which inputs will lead to the desired outcomes. The concept indeed applies directly to our pursuit of value.

Perhaps helpful for this concept is a story of one of the authors' childhood visits to the Ottawa Science Centre. He came upon one exhibit in which an undisclosed wooden shape was hidden under a larger wooden circle. Visitors to the exhibit couldn't directly observe which shape was hidden under the circle, but they could roll small balls under the larger piece of wood and watch them bounce back out in different directions. Through observation of these outcomes, hypotheses could be formed about which shape was hidden under the circle.

The Blue Line Imperative asks us to do the same in our pursuit of value. The mystery shape underneath the large circle is the set of value drivers that, if managed successfully, can take the company into the future in a healthy and sustainable way. Each roll of the ball offers us an observation from which inferences can be made. The more times we roll the ball, and the more observations we can make, the more likely we will arrive at an accurate interpretation of what's really going on underneath the big wooden circle.

We might call those who aspire to understand what is inside the black box or underneath the wooden circle, Bohrsians, based on Bohr's desire to visualize the structure of the atom. Those who are more comfortable with the idea that they will never be able to see inside the box we might call Heisenbergians, based on Heisenberg's recognition that a visualized mental model is not necessary for using observation to accurately describe probabilistically how changes in inputs will change outcomes. According to this camp, it is only the probabilistic model that will enable us to make the correct inferences about which inputs (value drivers) determine outcomes (indicators), and how.

Regardless of which view you prefer, the only important factor to remember is that we must focus on, and must strive to understand and manage, only probabilistic outcomes, that is, *expected* outcomes – rather than confusedly assuming that any given observed outcome is itself the objective we are after. It matters little whether you are a Bohrsian or a Heisenbergian. What matters is that in your company you establish and maintain a process by which insights can be enabled, understanding can be achieved, and value-oriented behaviors can be encouraged.

We know the blue line cannot be seen. So we need to roll metaphorical balls to give us data points that will allow us to learn what we need to do to make it go up. Because we can't see the blue line, it is easy to fall into the trap of watching the red line instead and then mistakenly asking questions like *"What is the share price doing today?"* or *"What level of earnings per share is the market expecting this quarter?"* Remember that such questions do not get to the heart of what your business is doing. They simply tell you what the market is reflecting back. Knowing the share price or this year's earnings does not tell you *how* it got that way. Listen to what physics is telling us. An observation is just an observation, not a reason. We must roll the balls and learn the lessons.

## Indicators, Data, and the Blue Line Firm

Broadly speaking, there are two types of data normally found in business. Some data tell us how well we are doing, and include the full range of metrics we normally think of when we use the term Key Performance Indicators (KPI). These data tend to be gathered on a regular, systematic basis. The other kind of data help us determine the value impact of decisions. These data are what we call decision-focused and are, therefore, more sporadic. We pose hypotheses about the relations between certain actions or behaviors and their value impact. Data are then gathered and tested with the intention of confirming or refuting these hypotheses.

The quality and usefulness of such data depends on the question or experiment used to trigger responses. Ask the wrong question and the resulting data are useless in guiding value-based decisions. To take an example from our own experience, senior management at a large European bank wanted to know whether they should cut back investment on their retail network. To gain insight regarding this question, they asked a large cross-section of customers, *"Would you like access to a branch on a regular basis, or Internet access only?"*

Nearly all of the respondents said that they would like access to a branch, and so the seemingly obvious decision was made to keep the retail network open. On the surface, the bank's reaction seems indeed reasonable.

But a more careful experiment would have crafted the question differently. Suppose keeping the retail branch open costs $100 per customer per year. The proper question isn't whether they want access to a physical site. Wouldn't everyone? The more important question is whether they are willing to pay for it. Posing this question might reveal that, while these customers would certainly prefer physical access to a branch, they aren't necessarily willing to pay $100 per year to get it. The value-based decision in this case was in fact to scale back the retail network. The bank realized this – a few years later.

Not only must we pose the right questions, but we must also be open to the answers. Imagine a scientist conducting an experiment. For genuine learning to occur, the experiment needs to be unbiased, rigorously constructed, and based on relevant and reliable data, but equally important, the scientist must be sufficiently open-minded and willing to incorporate any new insights that arise from the experiment.

It was recently revealed that there may be a flaw in Einstein's Special Theory of Relativity. Scientists in Europe had gathered data from three years' worth of experiments indicating that neutrinos in fact travel faster than the speed of light. It was possible, of course, that the experimental designs revealing this information were flawed, and that the resulting data was compromised. On the other hand, the data may have been correct, leading perhaps to Einstein's theory being superseded by something more accurate, in the same way that Einstein's work itself supplanted Newtonian physics.

Certain Einstein devotees, including those who had no direct access to the data nor the experimental design, quickly argued against the mere suggestion, stating unequivocally that the new data must be false. They revealed only a deep and abiding belief in the correctness of Einstein's theory; they were not indicating any willingness to accept new insights. (In their defense, an electrical fault was eventually identified as having led to inaccurate readings – Einstein's theory is intact, for now.) We must never assume that any previous information is gospel. Continuous learning comes only from designing our experiments objectively, asking the questions from a value perspective, and being open to the answers that arise from the balls we roll.

Blue line firms operate as an endless series of controlled experiments in the ongoing pursuit to identify their value drivers and deepen the understanding of what makes their business tick. This is the process that leads to value-based decisions, and it rests on the three pillars discussed earlier. Only if the organization has a culture of trust and fairness and a collective aim for learning can it succeed; and only if the experiments are designed from a blue line perspective will they yield the kind of results that can improve the organization's ability to deliver value to its customers.

Here's a basic example of what we mean. We think that the number of people in your company wearing orange socks might be an important indicator of whether you are managing the company for value, so we think you should try to get more people to wear orange socks. We understand that this may become a time-consuming and expensive effort, involving rebranding the company newsletter in orange, having a company-wide "Orange Week," and making sure the employee cafeteria serves oranges only as its fruit option, but we insist it's important.

There is only one logical response to our suggestion, of course. The response is not that the suggestion is ludicrous, though it may be. The response is that, before you listen to our suggestion or discard it, you need to assess and challenge the hypothesis, and thus determine whether the incidence of orange socks is really a value driver or whether we're completely out to lunch.

To do this, you must begin the statistical task of gathering data to test the hypothesis. This analysis is performed based on the relationship between a causal (or independent) variable, such as the percentage of employees wearing orange socks; and a dependent variable, such as profit margins, output per employee, or some other performance measure. If a carefully constructed test reveals no significant relationship between the two, you may conclude that wearing orange socks has nothing to do with value creation in your company. You may then conclude that the indicator "the percentage of our people who wear orange socks to work" is unrelated to performance, and therefore not an indicator worth tracking.

You might then look for a different driver – purple shirts perhaps, or making every other Friday Hawaiian shirt day – and repeat the process to see if you've landed on a real value driver this time. When you find one it is the time to think about how to guide actions and behaviors toward greater value.

All aspects of your business must be addressed in the same way as your response to our suggestion about orange socks and their relationship to value in your organization. Use indicators in the right way. Choose one, develop a hypothesis around it, gather data to test the hypothesis, and then use the results to help you learn what drives value in your company and what doesn't.

## Value-Based Decision-Making in Action

To see what genuine value-driven decision-making looks like, consider this example from a participant in one of our executive programs who had worked for many years at a different auto manufacturer but at the time of the program was working at Toyota. Here is his story:

> When I first started at Toyota, it seemed that I spent more of my time reporting on what I was managing than managing what I was managing. We were continuously feeding data into reports and databases, but it took me just a few weeks before I began to appreciate the value and capabilities of such a process.

As an example, suppose we are thinking of running an advertisement for a Corolla in a Seattle newspaper. At my old employer, we would have asked what that might cost and try to negotiate the lowest price possible before posing the question to the team of whether it was a good idea to run the ad. There were usually five people or so involved in the discussion, and we might've gone around asking each person's opinion and the reasons behind it. No one had any real data to support their opinion, so the only way to resolve differences was to apply a rule, such as "majority wins," or, more often, "the most senior guy present gets to decide."

How are things different at Toyota? We have a lot of useful data, so in such a case we might ask, "How many times have we run a comparable ad in the same paper for the Corolla, or a comparable car?" We can compare ads of the same size, on the same page, of the same newspaper, on the same day of the week or the same month. We then would ask, what was the impact on sales of the relevant model just after the ad was placed (one day later, two days, three days) and within 25 miles, 50 miles, and 75 miles of downtown Seattle? Because our decision support system continuously gathers these kinds of data, this information is available and it can be quickly analyzed. Also, we are able to estimate the margins earned from the cars sold in the days following the ads. With all of these data, we can then estimate the incremental cash contribution of the ad. Based on having run comparable ads 57 times over the past three years, we might estimate that the value of the ad is $3,000. If the cost of running the ad is less than this, we run it. If it is more than this, we don't.

At Toyota, anyone is welcome to offer their input, but not their opinion. To have any influence at all the input has to address the quality of the data or the analytical approach used. For example, if the data say that the expected value of the ad, based on the past data, is $3,000 and the cost is $3,400, someone might point out that the weather forecast is calling for a weekend of sunny weather, and the average of the past includes rainy days as well as sunny days. When we run the data again, this time excluding cases when it rained after the ad placement, we find that the expected value climbs to $4,200. Comparing this to the cost, $3,400, we see that it is now worthwhile running the ad.

This story demonstrates both sides of the Blue Line Imperative – the precedence of information over opinion, and the logic and analytical rigor that must be applied to it in order to achieve learning.

Because Toyota's learning and decision-making process is data driven, it does not require a hierarchy to resolve differences of opinion. People at Toyota may not be sitting around the table reciting the value equation (the expected future free cash flows discounted at the Opportunity Cost of Capital (OCC)), but they are certainly putting it into practice. They are rightly eschewing opinions and listening to what the data is telling them, because they well understand that their customers are the ultimate arbiters of whether they are creating value or not.

Google offers an equally compelling example of a value-driven company, where decision-making is never allowed to be influenced by personal bias. Tech writer Steven Levy notes that, by the time of its IPO in 2004, Google had grown to the size where, as he puts it, *"a company usually sets aside its loose structures and adopts well-established management structures."*[2]

Google was famous, among other things, for its flat, bottom-up style of management. But every time the head count doubled, the same question came up again: could this style of management actually scale? Although Google's founders, Larry Page and Sergey Brin, were convinced it could, middle management structures unavoidably crept in as the company blossomed. Eventually, the decision was made to formalize a new management position, "Associate Product Manager," or APM.

Each APM would take responsibility for coordinating activities among a group of engineers, a task that the big bosses at Google had finally conceded needed to be specifically assumed by someone. The APMs tend to be young technically-oriented people with senior management ambitions.

While this position at Google carries management responsibilities, APMs don't give orders. Their job is effectively to charm and persuade the engineers into a certain way of thinking. As one senior Google

---

[2] Steven Levy, *In the Plex: How Google Thinks, Works, and Shapes our Lives*, (New York: Simon & Schuster, 2011), pp. 157–158.

manager says, because the APMs work with people often much older and more experienced than them, *"They don't have the authority to say, 'Because I said so.' They need to gather data, lobby the team, and win them over by data."*[3] In short, data is at the center of everything.

The Internet has provided businesses with countless ways of running experiments to produce data that can help managers determine which pricing and promotion strategies work and which do not without bosses imposing their will from the top. In the same vein the speed and convenience of experimentation has ramped up by orders of magnitude compared to what existed even a few years before the Web exploded, allowing companies to run real-time experiments on their customers.

Google runs countless experiments every day, testing new algorithms, alternative web designs, and new products and services. Amazon likewise tests an ongoing series of hypotheses to assess potential new service features, alternative check-out processes, or new recommendation engines for books. The success of this approach, as always, depends on both the carefulness of the experimental design and the willingness to accept, and learn from the insights that result.

Tim Harford, known for his "undercover economist" column in the *Financial Times*, offers numerous other cases of companies espousing the Blue Line Imperative through experimentation and learning in his recent book, *Adapt: Why Success Always Starts With Failure.*[4] One example comes from a chain of craft and fabric stores, Jo-Ann Fabric and Craft Stores. The company started using its website as a laboratory, offering different customers a variety of special deals selected at random. Scads of data were produced, and any prior expectations or opinions about what would work and what wouldn't were disallowed – they would simply let the customers decide. One result of this objectivity was that a promotion which probably almost anyone would have considered weak – buy one sewing machine, get a second one at 20% off – was not only adopted but proved highly successful. Apparently, customers found that

---

[3]Ibid., p. 159.

[4]Tim Harford, *Adapt: Why Success Always Starts With Failure*, (New York: Farrar, Straus and Giroux, 2011), p. 235.

the prospect of saving 10% tempting enough to hunt for friends who might also want to buy a sewing machine.

Credit-card providers have long used similar approaches via junk mail, combining a variety of offers to often arbitrarily selected customers and then pushing hard on those that are shown to hit a nerve. As Harford explains, *"These experiments turbocharge the randomized trial method . . . by layering multiple randomizations on top of each other to generate a very rich set of data."*[5]

Supermarket chains likewise randomize their price offerings, shelf placement, and the coupons they send to customers with loyalty cards, to the same end. McDonald's has equipped some of its stores with devices that track ordering patterns and other data that allow researchers to model the impact on productivity and sales of variations in menus, restaurant design, and other factors.[6]

Let us remember that companies dedicated to the blue line, to value creation, to achieving true sustainable competitive advantage, do not just cast lines in the water and then celebrate only when they catch a fish. They pay as much attention to *why* they didn't catch one. Randomized experiments work when the business tolerates and encourages failure as the surest mode of learning. In the same way that scientists know the overwhelming majority of their experiments won't succeed, but that these failures will eventually provide the information necessary for success, companies properly oriented toward value do not expected a 100% hit rate in the experiments they conduct – but they do insist on a 100% dedication to learning.

Putting it in the language of opportunity cost, we might say that, for any experiment, real failure occurs only when the value of the learning is less than the cash-flow loss from failure. As long as the value of the learning is greater, the experiment is value-enhancing. This is what Thomas Edison meant when he said, *"If I find 10,000 ways something won't work, I haven't failed. I am not discouraged, because every wrong attempt discarded is just one more step forward."*[7]

---

[5] Ibid., p. 235.

[6] http://www.mckinseyquarterly.com/Are_you_ready_for_the_era_of_big_data_2864

[7] Quoted in Harford, p. 236.

# A Brief Digression Regarding Management Consultants

Before we proceed to the next chapter and a detailed discussion about properly calculating Net Present Value (NPV), allow us to make a few comments about consultants, since they are often giving companies advice about how to increase value.

As professors at a global business school, we are frequently exposed to both sides of the consulting business: those offering the services, our MBA graduates, and those who acquire their services, namely, the participants in our executive education programs. A common criticism leveled by our program participants goes something like this: *"Consultants come into our company, gather information from our people, analyze the information, then feed it back to us in a set of recommendations that we could easily have come up with on our own. Meanwhile, we pay them millions and have to spend a lot of time getting the data for them and answering their questions."*

Even more egregious in their eyes, they say, is that they often suspect the chief role of the consultant is to validate an important decision that management has already taken. In other words, the consultant is paid to say what the top guys want to hear. A final source of irritation, and perhaps the one heard most commonly in general business circles, is that the consulting firms rely heavily on young, inexperienced staff recently out of business school. What can an experienced corporate manager possibly learn from someone so green?

Despite all this, we would like to suggest that these consultants can add significant value for their clients by gathering data with care and integrity. The biggest reason for this is that, if nothing else, they analyze the data without preconceived assumptions, opinions, or beliefs. At the best consulting firms, this approach is drilled in early and often. They are taught in their training, and from early on-the-job experience, that value cannot be delivered to the client unless the data gathered are reliable, informative, and rigorously examined.

External consultants, coming in from outside as they are, bring no bias or personal agendas to the table. They lack the preconceived notions that get in the way of unbiased data-gathering and analysis. As people in the company make decisions and observe outcomes over time, some will gain genuine insights into the underlying drivers of success, though others will almost inevitably form, and try to disseminate, opinions about what matters and what doesn't. The observations of an impartial, external party can cut through the compromising influence of the latter scenario.

When decision-making is opinion-driven, the inevitable result is that the preconceived assumptions and beliefs held by managers prevent them from analyzing data rigorously or with integrity. Ironically, it is the very absence of experience and the perceived need to defend strongly-held opinions, combined with a thoughtful and rigorous approach to gathering and analyzing the data, that makes the role of consultants, potentially so valuable. As two former McKinsey consultants write, *"Facts compensate for a consultant's lack of experience and intuition relative to an executive with years of business experience. . . . Despite (or possibly because of) the power of facts, many businesspeople fear them."*[8]

To better understand how external consultants can liberate corporate managers from deeply entrenched opinions, consider the example of a well-known global bank. A senior executive told us that efforts to gather and analyze data within the bank had upended long-held assumptions about which customers (low-net-worth, high-net-worth, commercial, personal, etc.) used which products and services, and how much the different customer segments valued different services.

For example, it had been assumed for years that small corporate clients and higher-net-worth individuals are more likely to do their banking online and don't value the branch network. After years of holding onto, and making decisions based on, such assumptions with little genuine insight or data to back them up, the bank's top managers finally decided to hire a consulting firm to help determine which customer segments really did rely mostly on branches.

The analysis revealed that a significant percentage of the customers in the branches were in fact from the two groups mentioned, which forced the bank's management to re-think the inveterate assumptions and beliefs that had mistakenly driven major strategic decisions in the past. In short, they came to the uncomfortable conclusion that their strategies had been fundamentally flawed. Years of value destruction had resulted,

---

[8]Ethan M. Rasiel and Paul N. Friga, *The McKinsey Mind*, (New York: McGraw-Hill, 2002), p. 52.

as resources were allocated and pricing decisions made on the basis of beliefs that simply weren't true. Had the consultants never been recruited, the flawed assumptions, entrenched as they were, would have persisted, the blue line would have continued to fall, and the company would not have survived. We are not saying consultants are infallible, or even that they are always effective. We are saying that for value to be created, it can sometimes help to have an objective eye.

# CHAPTER ELEVEN

# GETTING NET PRESENT VALUE RIGHT

We have discussed the significance of the blue line and the reasons a value orientation is the only sustainable one. We have described the behaviors that characterize environments in which value creation or value destruction are most likely to occur. We have underscored specific elements of a blue line culture and the pillars of trust, fairness, and learning that must drive it. The question you may now be asking (or may have been asking throughout previous chapters) is: how to estimate the value of a business decision? It is, as we have said, impossible to know precisely the value of any decision, so we must apply a rigorous process for valuation that will merely produce better estimates. Remember that the Blue Line Imperative emphasizes the quality of the process, not the nature of the outcome. As Dwight D. Eisenhower famously said, *"I have always found that plans are useless but planning is indispensable."* The process of estimating the value of a business decision is indispensable, even if the estimate obtained is itself of limited use.

Recall also that a proper blue line valuation methodology is effective because it asks the right questions. All of your answers may be wrong, but don't let this bother you. You can't and won't see the blue line, so you can't and won't precisely forecast outcomes. The valuation is not intended to provide an answer, but rather to provoke relevant questions that in turn generate insight into, and a better understanding of, both sides of the value equation – the expected future free cash flows and the Opportunity Cost of Capital (OCC).

Performing a valuation in such a way that a decision-maker will have sufficient confidence to use it is no easy task. Those of you who prepare valuations or have done so in the past know well the situation in which a person of authority in your organization dismisses your recommendation and selects an alternative investment which the valuation suggests is value-destroying. There are many reasons for this – personal incentives being one obvious candidate – but perhaps the primary cause is a lack of confidence in the valuation model and the "assumptions" on which the forecast was built. Our goal in this chapter is to offer a framework based on the blue line principles and concepts we have discussed in order to help you both perform effective valuations and, just as important, communicate their implications with confidence and clarity to ensure that those in decision-making positions will listen.

Building a strong model is only the first step in the overall valuation exercise. Once you have your model, typically in the form of an Excel spreadsheet, the next challenge is to obtain a sensible, internally consistent scenario for the key value drivers that will determine your cash flow forecast. This process requires a thorough understanding of your business, which may be obtained only by analyzing available historical data and using it to build hypotheses for the underlying business drivers.

The next challenge concerns the *continuing* value that occurs after the explicit forecast period, typically three to ten years.

Finally, once we have estimates for the complete cash flow forecasts, we must incorporate the other side of the value equation – the discount rate required to convert the expected future cash flows into a present value equivalent. At this point we may finally perform the only comparison that matters: an assessment of whether the cash coming in is sufficient to compensate for the cash going out.

There are complications along the way that make the process additionally challenging for just about anybody who undertakes it. Certain elements of the process require some facility with mathematics; others require an understanding of the principles and tools of modern finance; and still others depend on a thorough knowledge of business strategy and economics. We will therefore approach this process via the following four steps:

1. The mechanics of discounting.
2. The mechanics of free cash flow.
3. Estimating an appropriate discount rate.
4. Understanding the economics of the business and its industry well enough to ensure that the results of the first three steps make sense.

Completing these steps with care and integrity is an ongoing yet essential challenge. Allow us to return once more to the two definitions of value: from the consumer's perspective, "Happiness"; and from the perspective of the businesses who exist to deliver that happiness, "The Expected Future Free Cash Flows Discounted at The Opportunity Cost of Capital." Here we are paying attention only to the second definition since, unlike the first, which is highly subjective and unique to each individual, it is determined by two objective components. Never forget that this definition does not ask for *your* expected future free cash flows, or for *your* OCC, but for *the* expected future free cash flows and *the* OCC. Your task in valuation is to estimate both. To understand the first, one must be able to grasp the economics of the business opportunity being valued. To understand the second, one must get finance. Only by achieving both of these tasks can one obtain a reasonable estimate of value for any business idea, project, or entity.

To better help you understand and entrench the blue line valuation process, let's go through a sample exercise involving the mechanics of discounting, a topic introduced in Chapter 4. Consider the following scenario:

You are buying a used car from your uncle, who agrees to let you pay the total cost of the car in two payments – one now, and the second in two years, when you expect your financial situation will have improved.

The first payment is to be $1,500 and the remaining payment $3,500. The current market rate for car loans is 6% per year. How much are you paying for the car in present value?

The answer is $4,615 = $1,500 + [$3,500/(1.06^2)]$. The first payment is not discounted because it is made in the present. The second payment will be made in the future and therefore must be converted to a present value equivalent. The logic here is that the expected future cash flow – $3,500 – is worth less in present value terms because the recipient has to wait to get the funds (and because of our mortality, might not even survive long enough to receive the second payment), and because of the possibility, however remote, that we will renege on our promise to pay or, for whatever reason, will be unable to pay.

When we calculate the present value of a future expected cash flow, such as the $3,500 in this example, we are essentially asking the question: how much would have had to be invested today, in another opportunity of similar risk, in order to have enough to pay the proposed amount on the date of that payment? In the above case, we are asking: how much needs to be invested today in an opportunity of similar risk in order to have $3,500 at the end of two years? If we begin with an amount, say, $1,000, and invest it in an opportunity of similar risk, according to the estimate above, the market would provide a return of 6% per year in such an opportunity. So at the end of the first year, we would have our investment plus a return of 6% on it, in other words: $1,000 \times (1 + 6\%) = $1,060$. Similarly investing this amount for the second year would give us a final amount of $1,123.60 (which is calculated as $1,000 \times (1 + 6\%)^2$).

Therefore, to calculate the present value, we work in reverse. For example how much would we need to invest today in order to have $110 at the end of one year when the OCC is 10%. To answer, we divide the amount we need by (1 + OCC), which in this simple example means the Present Value $= $110/(1 + 10\%) = $100$. For the car example above, if we are going to have to pay $3,500 in two years, then the present value of this is the amount we would have to invest in another opportunity of similar risk in order to have this amount of money at the end of two years. We calculate this as $3,500/(1 + 6\%)^2 = $3,115$. When added to the first payment of $1,500, this gives a total cost of $4,615 for the car.

Table 11.1: An illustration of free
cash flow

| | |
|---|---:|
| Revenues | 1,000 |
| COGS | −600 |
| SG&A | −100 |
| Depreciation | −100 |
| EBIT | 200 |
| Taxes on EBIT | −50 |
| NOPAT | 150 |
| + Depreciation | 100 |
| − Increase WCR | −10 |
| − Capex | −110 |
| Free Cash Flow | 130 |

## Mechanics and the Definition of Free Cash Flow

Table 11.1 above is the same one that appeared in Chapter 4 during our discussion of Free Cash Flow (FCF). The important point to recall when estimating FCF for decision-making is that *all* incremental cash flows impacted by the decision must be included, but *only* incremental cash flows impacted by the decision may be included. For example, any cash flows expected to occur independently of the decision are not relevant to that decision and should be excluded. Likewise, cash flows that have already occurred (also known as "sunk costs") should be excluded.

We have chosen to value a potential new business opportunity under consideration in early 2014 by a fictitious global tire and rubber manufacturer, referred to as "GlobalTire," which are based on the attributes of several larger players in the tire industry. The opportunity under consideration is a new line of agricultural tires for the North American market.[1]

---

[1] The example we provide contains numbers intended only for demonstration of our valuation methodology and do not reflect an actual project or decision undertaken by any global tire manufacturer.

We will need to forecast the expected future free cash flows and estimate the OCC to determine the appropriate discount rate for the forecasted cash flows.

Again, we accept that we can never really know the "true" expected future free cash flows, so we instead hope to apply a process that will provide us reliable estimates. To do this, we heed only data and logic, and when new data arrives, we incorporate it by following its logical implications. While ultimately our forecasted cash flows may still be poor estimates of the "true" expected future free cash flows, our ongoing learning over the course of repeated estimation efforts will increase both the likelihood that over time the estimates will improve, and that we will feel more confident in performing such forecasts for our business. In this way, the *process* will lead to a better understanding of the business we are trying to value and manage. Another way to say this is that we are attempting to do a valuation which is free of assumptions. When we say "free of assumptions," we simply mean that when asked *"why did you use that number?"* the answer cannot be *"I assumed it."* Instead, the answer must refer to a combination of data, e.g., *"it is the average of the past three years,"* and logic. As long as each number can be supported with an explanation that incorporates data and logic, and isn't *"I assumed it,"* then we have an assumption-free forecast. Note that this in no way ensures that the numbers are correct. We will still be quite sure that the numbers will all be proven wrong in the future. However, it does reduce bias in the forecasting process, as well as ensure the possibility of sharing and learning and improving on the forecasts through iterative application and by involving additional people to the effort.

Before diving in, we will need to consider many things: the industry; the business; the opportunity; the marketplace and current customer demand for the proposed product; the technology required; production and supply chain elements and costs; and the required investment in property, plant, and equipment. For the purposes of this exercise, let's assume we have done our homework on all these elements and we are well armed with data and research. We may now begin.

As we said, the opportunity under assessment is a new line of agricultural tires for the North American market. As scientists are increasingly aware of the negative impact of soil compaction caused by driving heavy equipment on farmland, the R&D function at GlobalTire has sought to address this problem by providing new tire technology which increases the tire footprint by over 25% while using the same-sized rims.

Various technological advances are employed to accomplish this feat, including new radial tire technology enabling greater sidewall flexibility, variable tire pressure (enabling farmers to travel on regular roads at high tire pressure but reduce the pressure to as low as 6 psi for farmland use), a flatter tread profile (which also extends service life), and many other innovations. The R&D folks at GlobalTire are justifiably proud of their long and successful tradition of tire innovations, which have improved the functionality, safety, durability, and fuel-efficiency of tires for many users, and they are keen to develop and deliver this new advance to the agricultural sector.

Unfortunately, while soil engineers and researchers are convinced of the benefits of reduced soil compaction to maintain healthy, fertile soil, many farmers are either unaware or unconvinced that the benefits out-weigh the costs. Therefore, Marketing is preparing promotional and educational materials to be delivered in various formats (adverts in agri-cultural publications, demonstrations at exhibitions and fairs, etc.), and the sales team will receive training and support to complement these efforts. Several advantages of the new tire are being highlighted in the materials in addition to reduced soil compaction, including: reduced likelihood of missed opportunities due to wet weather as the increased surface area (footprint) means farmers are able to work their land even in wet-weather conditions that might otherwise have been impossible; reduced downtime due to having vehicles "stuck" in the mud during wet weather; reduced "rutting" (the tire sinking into soft soil), which will improve yields and reduce the need for additional costly work to remove the ruts caused by normal tires; fuel efficiency, saving up to 25% on fuel in comparison to "standard" tires; and tire durability, extending the service life of the tire by one and a half years in comparison to previous-generation offerings. All of the R&D, marketing, sales, and general and administrative costs have been estimated according to the number of FTE (Full-Time Equivalent employees) in the various functions, as well as any other costs associated with them. For example, any outsourced market support would be included.

The line of tires, to be known as the UltraWide, is targeted at larger farm vehicles like combines and harvesters, used to harvest corn, wheat, soybeans, and the like. As such, the addressable market has been identified as farms of 500 acres or more that grow the appropriate crops. It is estimated that there are 160,000 farms in North America that meet the size requirements, with this number growing at 0.5% per year. In addition, it is estimated that 70% of these farms raise the relevant crops to be potential buyers.

Furthermore, the research gathered indicates that the typical farm owns two vehicles for which the new tires are appropriate, requiring four tires per vehicle. This results in a total market of approximately 896,000 units. Further market analysis has indicated an average replacement cycle of eight years (averaged between new equipment and new tires), and it is estimated that the new tire is likely to attain 14% market share of the combined Original Equipment (OE) and replacement tire market (growing steadily from an introductory market share of 3% in the first year to 14% in the fifth year after launch).

Finally, the tire has been priced at $3,500 (growing with inflation), which is at the high end of the spectrum to reflect the premium nature of the tire and the premium positioning of the GlobalTire brand.

The resulting sales forecast, which starts in 2013, is:

Table 11.2: Forecast sales for new agricultural tire business

| Sales Forecast | 2015 | 2016 | 2017 | 2018 | 2019 | 2020 | 2021 | 2022 | 2023 | 2024 |
|---|---|---|---|---|---|---|---|---|---|---|
| Farms generating > $500,000 sales | 160,000 | 160,800 | 161,604 | 162,412 | 163,224 | 164,040 | 164,860 | 165,685 | 166,513 | 167,346 |
| % crop-based relevant for new tire | 70% | 70% | 70% | 70% | 70% | 70% | 70% | 70% | 70% | 70% |
| # vehicles per farm relevant for new tire | 2.00 | 2.00 | 2.00 | 2.00 | 2.00 | 2.00 | 2.00 | 2.00 | 2.00 | 2.00 |
| # tires per vehicle | 4.00 | 4.00 | 4.00 | 4.00 | 4.00 | 4.00 | 4.00 | 4.00 | 4.00 | 4.00 |
| Potential market (units) | 896,000 | 900,480 | 904,982 | 909,507 | 914,055 | 918,625 | 923,218 | 927,834 | 932,474 | 937,136 |
| Replacement (vehicle or tire) cycle (years) | 8 | | | | | | | | | |
| Market share anticipated | 3.00% | 6.00% | 9.00% | 11.50% | 14.00% | 14.00% | 14.00% | 14.00% | 14.00% | 14.00% |
| Expected sales (units) | 3,360 | 6,754 | 10,181 | 13,074 | 15,996 | 16,076 | 16,156 | 16,237 | 16,318 | 16,400 |
| Price per unit | 3,500 | 3,579 | 3,659 | 3,742 | 3,826 | 3,912 | 4,000 | 4,090 | 4,182 | 4,276 |
| Sales forecast | 11,760,000 | 24,169,446 | 37,255,237 | 48,918,377 | 61,197,209 | 62,887,017 | 64,623,485 | 66,407,901 | 68,241,589 | 70,125,910 |

In order to deliver these new tires, the production engineers, working together with the product development and R&D teams, have designed a new cutting-edge manufacturing facility designed with the aid of ongoing efforts to introduce "lean manufacturing" and "Six Sigma" processes that will ensure rapid process improvement following the plant opening. While the unanticipated always lurks around the corner, the production team is sure that their new lean and continuous-improvement methods will help iron out the production wrinkles in record time. The extra investment, bringing the total cost of the facility to $30,000,000, will enable additional flexibility and more rapid and error-free (at least, reduced error-rate) ramp-up of production volumes to meet the anticipated demand as the new tire grabs market share over the five years to come.

To maintain quality and meet volume demand, it is anticipated that ongoing investment in plant and equipment, as well as in computers and other equipment for the growing office staff, will be a further 5% of the upfront investment each year for five years. That will be followed by 10% each year thereafter, as greater refurbishment and equipment replacement becomes necessary in anticipation of substantial technological improvements.

The overall costs associated with the new business are summarized below:

Table 11.3: Agricultural tire business cost summary

| Cost Estimates | | | | |
| --- | --- | --- | --- | --- |
| New plant construction | 30,000,000 | ← initially $30,000,000 then refurbishment of 5% in each year for 5 years and 10% each year thereafter | 5% | 10% |
| Raw materials for production per unit | 1,050 | | | |
| Energy for production/ transportation per unit | 315 | | | |

Table 11.3: (*Continued*)

| Cost Estimates | | | |
|---|---|---|---|
| Labour/other direct costs + Transportation (per unit) | 945 | | |
| General and administrative staff | 24 | ← initially 24 people at $90,000 per person total cost, growing to 36 people after 5 years | 36 |
| General and administrative staff salary/benefits cost | 90,000 | | |
| R&D staff | 8 | ← initially 8 people at $120,000 per person total cost, growing to 12 people after 5 years | 12 |
| R&D average salary/ benefits cost | 120,000 | | |
| Sales and Marketing (non-staff) | 1,500,000 | | 2,250,000 |
| Sales and Marketing staff/outsourced | 120 | ← initially $1,500,000 ad budget + 120 people on commission avg $12,500 per person, growing to $2,250,000 and 180 people after 5 years | 180 |
| Sales commission | 12,500 | | |

(*Continued*)

### Table 11.3: (*Continued*)

| Cost Estimates | | |
|---|---|---|
| Debt as % new investments | 20% | |
| ST debt as % LT debt | 10% | |
| Interest rate on the debt | 5.7% | |
| LT debt repaid per year (% previous year) | 5.0% | |
| Tax rate | 35.0% | |
| Cash as % revenue | 2.0% | |
| Accounts receivable (days sales outstanding) | 67.5 | ← Average of past 3 years |
| Inventory turnover (COGS/Inventory) | 3.1 | ← Average of past 3 years |
| Prepaid expense (% G&A) | 15.0% | ← leases and insurance |
| Accounts payable (days of COGS) | 51.6 | ← Average of past 3 years |

As described, the costs are based on a combination of internal experience and externally sourced data. Input quantities for raw materials have been estimated according to quantity estimates for the various materials (natural and synthetic rubber, carbon black, silica, nylon, steel, rayon, oil, resins, and other chemicals), used in the production of the tire according to the specifications of the R&D and design engineers, with the prices for each input determined from current market prices. Most of these materials are traded in active markets where the prices are readily observable. For example, natural rubber averaged $4.60/kg in 2011, butadiene (used in production of synthetic rubber) averaged $2.91/kg in 2011, and carbon black averaged $850/ton.[2]

---

[2] Sources: Global Tire Annual Results, February 10, 2012 and Alibaba.com

To estimate the costs of manufacturing and transportation, we reference internal estimates on the number of FTE required by manufacturing activity. This includes productivity metrics, employee attrition and training, as well as energy cost based upon similar production of similar tires using similar processes in one of the other 17 existing specialty tire plants operating around the globe; and in addition to this, logistics, distribution, and transportation costs based on existing numbers for North America. The result is manufacturing and transportation costs averaging approximately 68% of selling price (based on existing GlobalTire performance). However, for premium products, the average is lower, at approximately 65–66%, which is the estimate we will therefore use in the forecast. These Costs of Goods Sold (COGS) are broken into three sub-categories of raw materials, energy, and labor and other direct costs and are estimated based on historical data, at 30%, 9%, and 27% of selling price, respectively.

The remaining costs are divided into three categories: General and Administrative (G&A), Research and Development (R&D), and Marketing and Sales (M&S). For each category, we make use of evidence from similar projects in the agricultural segment where the number of people working in each area, as well as the productivity and activities, are easily known. With these data, we can estimate the number of people actually needed to perform the tasks which comprise G&A, such as information technology (IT), human resources (HR), and accounting and finance. For this project, the number is estimated to be 24 FTE at launch, rising to 36 when full capacity is reached in five years.[3]

We then need an estimate of the current fully loaded cost per FTE (including salary, benefits, vacation/sick time, real estate, and equipment including desks, computers, software, and so forth). Salaries vary substantially across G&A staff, but the average salary is estimated at $58,000, and the additional charges required for the fully loaded cost are estimated at $32,000, giving us an average fully loaded cost per FTE of $90,000.

We follow the same process to estimate that we will need 8 people in the R&D function devoted full-time (or equivalent) to the project, and that this number will rise to 12 people as the product reaches its market potential and further refinements and innovations become necessary.

---

[3]The estimate of five years is based on the market share evolution anticipated by the marketing department, which in turn is based on prior experience.

Based on existing salaries and costs, the average fully loaded cost of the staff in the R&D function is $120,000.

According to recent experience, the fully loaded costs of all staff are expected to rise only with inflation during the life of the project, even though their productivity should increase with adoption of new technologies.

Finally, to estimate costs for marketing and sales, we must speak with the people in those departments. It turns out they are excited by recent growth in the agricultural sector and have developed strong brand awareness in the industry through a concerted, mixed approach to marketing and promotion using various media and incentives for resellers. They intend to ensure customer pull by using knowledge gained from recent experience in each channel, including online/social networking, physical presence at agricultural shows/fairs, targeted adverts in appropriate publications and websites, and promotional materials and sales visits to the dealer network – both tire dealers and vehicle dealers in the relevant agricultural communities.

They are also targeting customer push through both their salesforce and relationships with vehicle manufacturers (the OE manufacturers) to ensure the new tires are on as many vehicles as possible that fit the criteria. To fulfill this substantial goal, approximately 10% of the existing sales force for agricultural tires in North America will be mobilized, trained, and compensated for selling the new tire, and this percentage will rise to 15% when the tire is expecting to be achieving its full market potential five years from now.

With an estimated additional cost per existing sales person of $12,500 associated directly to this new effort, and with a total of 1,200 existing (salaried and sub-contracted) sales staff across North America, the basic marketing and sales expenses associated with this project are estimated. In addition, the marketing team plans to spend a total of $1,500,000 on marketing in the first year, and they intend to raise this spend to $2,250,000 by the fifth year after launch.

Having assembled these revenue and cost estimates, we are not quite able to build a forecast of the income statement, but we are getting close. There are just a few missing elements to fill in: Depreciation – determined by the investment pattern and the depreciation schedule of the assets acquired from those investments; Interest – which depends upon debt financing and the interest charged on it; and Taxes – which depend upon profits earned and the applicable income tax rate.

Let's tackle Depreciation first. According to the capital investment agenda described above, the upfront investment of $30 million in 2014, will be followed by investments of $1.5 million per year until 2019, and $3 million per year thereafter. The depreciation associated with these investments is estimated using a straight-line schedule beginning the year of the original investment.

For this reason, the initial $30 million investment will generate a depreciation expense of $3 million for 2013 and each of the subsequent 9 years, as shown on the income statement below. The additional capital investment of $1.5 million in 2015 will generate further depreciation starting in 2016 of $0.15 million per year for the following decade. The total depreciation in any given year is the sum of these depreciation expenses for any past investment still being depreciated.

To estimate Interest, we must understand how the project will be financed. Given GlobalTire's average debt levels of the recent past, we estimate that debt provides 20% of the investment capital needed (in this case, 20% of the $30 million initially, plus any additional investment during the life of the project), and further that GlobalTire will make use of this debt capacity accordingly to obtain benefits that will reduce their tax liability. The debt is to be sourced as GlobalTire's total debt has been, with 90% financed long term and the remaining 10% financed short term. Based on anticipated maturity schedules for the long-term debt, we estimate that GlobalTire will pay off 5% of the outstanding long-term debt each year (and all of the short-term debt), to be replaced with new debt depending upon ongoing investments and debt capacity, again at 20% of new investments. Finally, we have estimated an interest rate of 5.7% based on the rate at which GlobalTire has obtained financing recently and the current macroeconomic situation globally and in the US.

The final item on the income statement is Taxes. As this business is to be conducted in North America with GlobalTire's headquarters located in Greenville, South Carolina, we estimate that 35% of future pre-tax income will be paid as taxes.

All of these considerations, taken together, produce the following income statements:

Table 11.4: Forecast income statements for agricultural tire business

| Income Statement | 2015 | 2016 | 2017 | 2018 | 2019 | 2020 | 2021 | 2022 | 2023 | 2024 |
|---|---|---|---|---|---|---|---|---|---|---|
| $ thousands | | | | | | | | | | |
| Sales | 11,760 | 24,169 | 37,255 | 48,918 | 61,197 | 62,887 | 64,623 | 66,408 | 68,242 | 70,126 |
| COGS | (7,762) | (15,952) | (24,588) | (32,286) | (40,390) | (41,505) | (42,651) | (43,829) | (45,039) | (46,283) |
| General and administrative | (2,160) | (2,485) | (2,823) | (3,175) | (3,542) | (3,621) | (3,703) | (3,786) | (3,871) | (3,958) |
| R&D | (960) | (1,104) | (1,255) | (1,411) | (1,574) | (1,609) | (1,646) | (1,683) | (1,721) | (1,759) |
| Sales & marketing | (3,000) | (3,413) | (3,835) | (4,267) | (4,709) | (4,815) | (4,924) | (5,035) | (5,148) | (5,264) |
| Depreciation | (3,000) | (3,150) | (3,300) | (3,450) | (3,600) | (3,750) | (4,050) | (4,350) | (4,650) | (4,950) |
| EBIT | (5,122) | (1,934) | 1,454 | 4,329 | 7,382 | 7,585 | 7,650 | 7,725 | 7,813 | 7,912 |
| Interest | (340) | (340) | (340) | (340) | (340) | (340) | (357) | (373) | (388) | (403) |
| EBT | (5,461) | (2,274) | 1,114 | 3,989 | 7,042 | 7,246 | 7,293 | 7,353 | 7,424 | 7,509 |
| Taxes on EBT | 1,911 | 796 | (390) | (1,396) | (2,465) | (2,536) | (2,553) | (2,573) | (2,599) | (2,628) |
| EAT | (3,550) | (1,478) | 724 | 2,593 | 4,578 | 4,710 | 4,741 | 4,779 | 4,826 | 4,881 |

While producing these income statements was an important step – and far from an easy one – we are not yet ready to forecast the first part of the value equation, The Expected Future Free Cash Flows. We know the new manufacturing facility will cost $30 million plus ongoing refurbishments, but this is not the only investment required to get this project underway and keep it running.

The plant won't function without inputs, so there must be investment in *raw materials inventory* which will then be applied to the production process. This process will create *work-in-progress inventory* – the combination of raw materials and production costs that are investments needed to obtain partially transformed materials used during production. Upon completion of the production phase, we will have *finished goods inventory*, which reflects the investment needed to fully transform the raw materials into their final form.

These combined inventories (raw materials, work-in-progress, and finished goods) represent significant potential investment, and the cost of this investment must be reflected in our estimate of The Expected Future Free Cash Flows, since it is going to show up on the balance sheet, and any additional year-by-year investment in inventories will appear as increases (or decreases if we are disinvesting) in this item.

The agricultural tire project will have an inventory turnover cycle similar to that of any similar project within GlobalTire. Therefore the inventory turnover, measured as COGS/Inventory, is expected to average 3.1, the company average over the past three years. This is substantially below the industry average of 4.5, but it is nonetheless expected for the new project because of the anticipated application of established GlobalTire management practices.

Other important figures to determine are Accounts Receivable, Prepaid Expenses, and Accounts Payable, because the terms and conditions for these items are critical to the timing of cash flow. Revenues are "booked" when invoiced and/or delivered, but the actual cash may not be received for several weeks or even months thereafter depending on a number of factors, including the agreed terms with the customer.

Accounts Payable reflect the terms and conditions agreed with the company's suppliers – they deliver GlobalTire products and services, and GlobalTire agrees to pay them for doing so. To make its new tire, the company needs to buy raw materials, pay production costs, and spend money to warehouse and distribute them. However, those working in business will know that all of these expenditures often precede the actual cash payment from sales of the finished goods. So the respective timing of outlays and inflows represented by Accounts Payable and Accounts Receivable and Prepaid Expenses must be forecast accurately, and their impact on FCF accordingly reflected.

Regarding Accounts Receivable, GlobalTire, like any stable company, attempts to manage its customer relationships effectively to maintain their loyalty and satisfaction while also ensuring they pay for the products. Payment terms in the tire industry are relatively favorable to customers, with an overall average of 60+ days of sales outstanding as receivables at any point in time. On this dimension GlobalTire is an average performer, with an average over the past three years of 67.5 days. We apply this average to the proposed project. For their Prepaid Expenses, the group average is that approximately 15% of G&A expenditures needs to be paid in advance (leases and insurance, primarily).

Concerning Accounts Payable, which enables us to take delivery from suppliers in advance of having to pay, thus freeing cash for other uses, the industry average for the past three years is approximately 70 days. GlobalTire on this front is faster than its competitors, paying its suppliers in an average of 51.6 days, which is the estimate we will apply to the new tire.

Having populated both the income statement and balance sheet, we are now able to develop the forecast cash flow statements for the proposed new agricultural tire business.

Note that the forecast cash flow statement is not quite the same as the forecast for *free* cash flow, shown below. The cash flow statement is useful in helping us understand and anticipate the basic flow of money into and out of the project. It also helps us arrange the needed sources of funding to avoid surprises. But it is The Expected Future Free Cash Flow – the first half of the value equation – that we will use to value the project.

In our forecast, the impact of any financing on sales and costs has been incorporated, since these figures will be similar to those of the past, and our ratios have been accordingly aligned. However, the impact on taxes has not yet been incorporated and does not show up in the mechanics of the definition of FCF, as shown in Table 11.7.

Table 11.5: Balance sheet for agricultural tire business

| Balance Sheet | 2014 | 2015 | 2016 | 2017 | 2018 | 2019 | 2020 | 2021 | 2022 | 2023 | 2024 |
|---|---|---|---|---|---|---|---|---|---|---|---|
| $ thousands | | | | | | | | | | | |
| **Assets** | | | | | | | | | | | |
| Cash | 235 | 235 | 483 | 745 | 978 | 1,224 | 1,258 | 1,292 | 1,328 | 1,365 | 1,403 |
| Accounts receivable | 706 | 2,175 | 4,470 | 6,890 | 9,047 | 11,317 | 11,630 | 11,951 | 12,281 | 12,620 | 12,968 |
| Inventories | | 2,504 | 5,146 | 7,932 | 10,415 | 13,029 | 13,389 | 13,759 | 14,138 | 14,529 | 14,930 |
| Prepaid expense | 324 | 324 | 373 | 423 | 476 | 531 | 543 | 555 | 568 | 581 | 594 |
| Property, plant, and equipment | 30,000 | 31,500 | 33,000 | 34,500 | 36,000 | 37,500 | 40,500 | 43,500 | 46,500 | 49,500 | 52,500 |
| Cumulated depreciation | | (3,000) | (6,150) | (9,450) | (12,900) | (16,500) | (20,250) | (24,300) | (28,650) | (33,300) | (38,250) |
| Net property, plant, and equipment | 30,000 | 28,500 | 26,850 | 25,050 | 23,100 | 21,000 | 20,250 | 19,200 | 17,850 | 16,200 | 14,250 |
| **Total Assets** | **31,265** | **33,738** | **37,322** | **41,040** | **44,016** | **47,102** | **47,070** | **46,757** | **46,165** | **45,294** | **44,145** |
| **Liabilities + Shareholders' Equity** | | | | | | | | | | | |
| Short-term debt | 545 | 545 | 545 | 545 | 545 | 545 | 573 | 599 | 623 | 647 | 669 |
| Accounts payable | – | 1,097 | 2,255 | 3,476 | 4,564 | 5,710 | 5,868 | 6,030 | 6,196 | 6,367 | 6,543 |
| Long-term debt | 5,455 | 5,455 | 5,455 | 5,455 | 5,455 | 5,455 | 5,727 | 5,986 | 6,233 | 6,466 | 6,688 |
| Equity | 25,265 | 26,640 | 29,066 | 31,564 | 33,452 | 35,392 | 34,902 | 34,143 | 33,114 | 31,814 | 30,244 |
| **Total Liabilities & Shareholders' Equity** | **31,265** | **33,738** | **37,322** | **41,040** | **44,016** | **47,102** | **47,070** | **46,757** | **46,165** | **45,294** | **44,145** |

## Table 11.6: Cash flow statements for agricultural tire business

| Cash Flow Statement | 2014 | 2015 | 2016 | 2017 | 2018 | 2019 | 2020 | 2021 | 2022 | 2023 | 2024 |
|---|---|---|---|---|---|---|---|---|---|---|---|
| **$ thousands** | | | | | | | | | | | |
| **Operating Activities** | | | | | | | | | | | |
| Opening cash balance | – | 235 | 914 | 1,842 | 2,783 | 3,695 | 4,620 | 5,333 | 6,081 | 6,862 | 7,675 |
| EAT | – | (3,550) | (1,478) | 724 | 2,593 | 4,578 | 4,710 | 4,741 | 4,779 | 4,826 | 4,881 |
| Interest | – | 340 | 340 | 340 | 340 | 340 | 340 | 357 | 373 | 388 | 403 |
| Depreciation | – | 3,000 | 3,150 | 3,300 | 3,450 | 3,600 | 3,750 | 4,050 | 4,350 | 4,650 | 4,950 |
| Change in WCR | (1,030) | (2,876) | (3,828) | (4,036) | (3,605) | (3,794) | (527) | (541) | (556) | (571) | (587) |
| Net Operating Casf Flow | (1,030) | (3,086) | (1,816) | 328 | 2,778 | 4,723 | 8,273 | 8,606 | 8,946 | 9,293 | 9,647 |
| **Investing Activities** | | | | | | | | | | | |
| Sale of fixed assets | | | | | | | | | | | |
| Capital expenditures | (30,000) | (1,500) | (1,500) | (1,500) | (1,500) | (1,500) | (3,000) | (3,000) | (3,000) | (3,000) | (3,000) |
| **Financing Activities** | | | | | | | | | | | |
| Increase (decrease) in ST Debt | 545 | – | – | – | – | – | 27 | 26 | 25 | 23 | 22 |
| Increase (decrease) in LT Debt | 5,455 | – | – | – | – | – | 273 | 259 | 246 | 234 | 222 |
| Interest payments | – | 340 | 340 | 340 | 340 | 340 | 340 | 357 | 373 | 388 | 403 |
| Equity infusion (dividend) | 25,265 | 4,925 | 3,904 | 1,773 | (705) | (2,638) | (5,199) | (5,500) | (5,808) | (6,125) | (6,451) |
| Dividend payments | – | – | – | – | – | – | – | – | – | – | – |
| Total cash flow from financing | 31,265 | 5,265 | 4,244 | 2,113 | (365) | (2,298) | (4,560) | (4,858) | (5,165) | (5,480) | (5,804) |
| Closing cash balance | 235 | 914 | 1,842 | 2,783 | 3,695 | 4,620 | 5,333 | 6,081 | 6,862 | 7,675 | 8,518 |

## Table 11.7: Forecast of free cash flow for agricultural tire business

| Free Cash Flow | 2014 | 2015 | 2016 | 2017 | 2018 | 2019 | 2020 | 2021 | 2022 | 2023 | 2024 |
|---|---|---|---|---|---|---|---|---|---|---|---|
| $ thousands | | | | | | | | | | | |
| Revenues | | 11,760 | 24,169 | 37,255 | 48,918 | 61,197 | 62,887 | 64,623 | 66,408 | 68,242 | 70,126 |
| COGS | | (7,762) | (15,952) | (24,588) | (32,286) | (40,390) | (41,505) | (42,651) | (43,829) | (45,039) | (46,283) |
| SG&A | | (6,120) | (7,002) | (7,913) | (8,854) | (9,825) | (10,046) | (10,272) | (10,503) | (10,740) | (10,981) |
| Depreciation | | (3,000) | (3,150) | (3,300) | (3,450) | (3,600) | (3,750) | (4,050) | (4,350) | (4,650) | (4,950) |
| EBIT | | (5,122) | (1,934) | 1,454 | 4,329 | 7,382 | 7,585 | 7,650 | 7,725 | 7,813 | 7,912 |
| Taxes on EBIT | | 1,793 | 677 | (509) | (1,515) | (2,584) | (2,655) | (2,677) | (2,704) | (2,734) | (2,769) |
| NOPAT | | (3,329) | (1,257) | 945 | 2,814 | 4,798 | 4,931 | 4,972 | 5,022 | 5,078 | 5,143 |
| Add back Depreciation | – | 3,000 | 3,150 | 3,300 | 3,450 | 3,600 | 3,750 | 4,050 | 4,350 | 4,650 | 4,950 |
| Subtract increase in WCR | (1,030) | (2,876) | (3,828) | (4,036) | (3,605) | (3,794) | (527) | (541) | (556) | (571) | (587) |
| Subtract Capex | (30,000) | (1,500) | (1,500) | (1,500) | (1,500) | (1,500) | (3,000) | (3,000) | (3,000) | (3,000) | (3,000) |
| FREE CASH FLOW | (31,030) | (4,705) | (3,435) | (1,291) | 1,159 | 3,104 | 5,154 | 5,481 | 5,816 | 6,157 | 6,506 |

Remember that it is a common error to believe that the impact of interest on value is via increased risk. This is not the case. The impact of interest on value concerns the reduced tax cash flows available to send to the government. When the company reduces its taxes, it has more cash available for investors.

To incorporate this impact, we can either change the discount rate using the Weighted-Average Cost of Capital (WACC) formula to simultaneously reflect the riskiness of the business, assumed to be the same as the average business risk for the tire industry, as well as the tax impact of the interest payments on the debt financing – or we can use a technique called the Adjusted Present Value (APV) method to separate the two elements. We demonstrate the outcome for both methods below.

Note that we have added the year 2025 in order to incorporate the expected ongoing benefit of the business beyond just the 10 years of our forecast. We have made use of the growing perpetuity formula and the discussion from Chapter 7. Recall that growth in itself cannot be described as value creation, but the continuation of the business, and the corresponding cash flows it generates, are of substantial value. This needs to be reflected in our estimation of the overall value of the new business opportunity.

As you can see in Table 11.8, whether we use the APV or the WACC method, which give similar, though not identical, answers, because of slight differences in the discounting of the estimated cash flows, the proposed project appears to yield a positive NPV. If we are value based and data driven, then we must conclude that it is the right move to proceed. Right?

Not so fast. We have captured one side of the value equation, The Expected Future Free Cash Flows, but we have not yet explained how we determined the discount rate which was used to convert the expected future free cash flows to present value in the table above. We turn to this discussion now.

## The Mechanics of Estimating an Appropriate Discount Rate

The other side of the value equation is, of course, the OCC. Without it, our estimate of the proposed investment's Net Present Value (NPV) will be worthless. Remember from Chapter 3 that the OCC is determined by nature, and that to properly estimate it, we need insight into nature's perspective on its two determining factors: first, the required return for

## Table 11.8: Valuation using APV and WACC methods

| Valuation: APV & WACC | 2014 | 2015 | 2016 | 2017 | 2018 | 2019 | 2020 | 2021 | 2022 | 2023 | 2024 | 2025 |
|---|---|---|---|---|---|---|---|---|---|---|---|---|
| **$ thousands** | | | | | | | | | | | | |
| Free cash flows | (31,030) | (4,705) | (3,435) | (1,291) | 1,159 | 3,104 | 5,154 | 5,481 | 5,816 | 6,157 | 6,506 | |
| Interest tax shields | | 119 | 119 | 119 | 119 | 119 | 119 | 125 | 130 | 136 | 141 | |
| Final year growth rate (2022–2023) | 4.0% | | | | | | | | | | | |
| Growth rate after 2023 | 4.0% | | | | | | | | | | | |
| **APV Method** | | | | | | | | | | | | |
| Discount rate for free cash flows | 9.4% | | | | | | | | | | | |
| Unlevered OCC discount factor | 1.00 | 0.91 | 0.84 | 0.76 | 0.70 | 0.64 | 0.58 | 0.53 | 0.49 | 0.45 | 0.41 | 0.37 |
| Discount rate for interest tax shields | 5.7% | | | | | | | | | | | |
| Debt OCC discount factor | 1.00 | 0.95 | 0.90 | 0.85 | 0.80 | 0.76 | 0.72 | 0.68 | 0.64 | 0.61 | 0.58 | 0.55 |
| PV (FCF)@ unlevered OCC | (31,030) | (4,302) | (2,872) | (987) | 810 | 1,984 | 3,012 | 2,929 | 2,841 | 2,750 | 2,657 | 21,321 |
| PV(Int tax shields) @ debt OCC | – | 113 | 106 | 101 | 95 | 90 | 85 | 85 | 84 | 83 | 81 | 1,020 |
| **NPV (APV Method)** | **1,057** | | | | | | | | | | | |
| **WACC Method** | | | | | | | | | | | | |
| WACC | 9.0% | | | | | | | | | | | |
| WACC discount factor by year | 1.00 | 0.92 | 0.84 | 0.77 | 0.71 | 0.65 | 0.59 | 0.55 | 0.50 | 0.46 | 0.42 | 0.39 |
| PV (FCF)@ WACC | (31,030) | (4,315) | (2,889) | (995) | 820 | 2,014 | 3,066 | 2,990 | 2,910 | 2,825 | 2,737 | 22,825 |
| **NPV (WACC Method)** | **958** | | | | | | | | | | | |

simply waiting (often referred to as the "time value of money"), and second, the required return to compensate for exposure to risk.

Recall that the OCC is *not* the Cost of Funding (COF), but that we are forced to use whatever observable data we have to obtain the estimates we need – so for our current purposes, we will indeed use the COF to estimate the OCC. It bears repeating that the process of estimating the OCC is far more important than the estimate itself.

Though we are using the COF as a means of estimating the OCC, it is paramount that the difference between them in application is understood. Many books on finance and valuation point out the conceptual distinction between COF and OCC but fail to take the next step of providing a methodology for applying them to an estimation of value. This is part of the reason for the widespread confusion about valuation. This confusion is partly due to the definition of the WACC formula and with the way it is typically used. Therefore, we will explain the inputs to the OCC and describe the method through which those inputs are reflected in the WACC and that will help to clarify why we use the WACC formula the way we do in practice.

Let's return for a moment to our definition of investing as moving energy through time and of the OCC as the required return for this effort. Consider the first determinant of the OCC, the required return for waiting (the "Time Value of Money"), estimated by the risk-free rate as described above. We know that nature will neither provide an infinitely positive return nor accept an infinitely negative one. Therefore, the number is determined and finite.

But what is the number? It is the rate nature will pay for the movement of energy into the future when there is no risk with respect to getting that energy back later (apart from the possibility of your not being alive to receive it). In most finance textbooks, you will read that the risk-free rate is the government bond rate or something comparable. We will try a different tack. Let us consider the act of moving energy into the future and see if we can actually observe the return earned on this effort.

To accomplish this, we would need to perform a large experiment involving many thousands, or ideally millions, or better yet, trillions, of smaller individual experiments in which we attempt to move energy into the future and then estimate the average return, yielding a reasonably good estimate of nature's risk-free rate.

One difficulty we will encounter is that, when individual economic agents are almost entirely focused on consumption, that is survival, and only a very few – the ruling class, shall we say – are able to enjoy the luxury of investment, we do not obtain reliable data. Indeed, whenever there is a ruling class, it tends to get rich by simply exploiting and stealing from the rest of us. They are consuming rather than investing. Our experiment therefore requires a significant population, of which a significant percentage must be able to individually determine how to apply themselves and move energy through time through entrepreneurial endeavors and without a clear ruling class able to exploit large numbers of them, since we know in that scenario the society quickly turns consumption-oriented rather than investment-oriented.

The example we used in Chapter 3 to demonstrate this point was the US economy from 1870 to 2009 – a period during which individual citizens were remarkably free to pursue their dreams as entrepreneurs with minimal interference from the state or other bodies. In addition they were, at the time, equally free to pursue their own consumer dreams by buying whatever they wished, whenever they wished, again with little to no intrusion or restriction. In this environment, the average return earned on efforts to move energy through time – in this case represented by GDP per capita – was 2%, a figure that remained startlingly consistent. This number is the outcome of trillions of individual efforts to move energy into the future through entrepreneurship and innovation. It is the aggregate outcome of a highly competitive environment in which each individual economic agent is relatively free to both produce and consume as he or she wishes. As such, it provides us with the best single estimate of nature's risk-free rate.

We could consider an alternative approach to estimating the same OCC. Imagine a highly competitive and transparent marketplace in which all participants agree that there will be no risk of default. The traditional example of such a security was a government bond but, by 2011, it had become abundantly clear to most observers that no government bond is truly risk-free, even in "rich" countries like Spain and Italy. Nevertheless some government bonds have a relatively low risk of default, and some, such as the US, Canada, Germany, Australia, Sweden, The Netherlands, and Singapore among others, have such a low risk of default we recognize that the returns demanded on these when determined in a large, liquid, transparent, and highly competitive market are reasonable esti-mates of the risk-free rate, and are themselves examples of securities referred to as "risk-free."

Suppose such a "risk-free" security will pay $100 a year from now. We could simply offer the security for sale in this highly competitive market and then see how it ends up priced by the participants. Initially, investors will try to buy it at the lowest possible price in order to generate the highest possible return – for example, paying $1 for the investment and thus receiving a return of 9,900%. But, due to the nature of competition, other investors will quickly jump in, preventing the first group from real-izing so generous a return. These new investors will bid up the price, say to $2, driving the return on the investment down to a mere 4,900%.

This process will continue until the point where no amount of addi-tional transparency, competitiveness, or new investors can impact the price any longer and it is at this point that we will be observing nature at work. Henceforth the price will simply bounce around its mean, some-times moving slightly up, other times slightly down.

Suppose this leveling off happens at a price of $98.04, for an implied risk-free rate of 2%. It would not be accurate to say that investors are happy to receive this 2%. Each of them would obviously rather pay a lower price and receive a higher return. The only reason they accept the 2% return is because it's what is available.

A fully transparent and competitive marketplace with many partici-pants does not reveal what humans want. It reveals what they will accept. Millions of humans acting competitively reveal nature. Rather than waiting 139 years for data on the actual return earned on efforts to move energy through time in the form of economic development in a competitive and free environment – the US GDP per capita from 1870 to 2009 – we can simply throw a risk-free security into a highly active, liquid, transparent, competitive marketplace and see how it gets priced nearly instantane-ously. The capital market provides exactly the insights into nature that we are looking for. As a proxy for a risk-free asset, we select the bonds of an entity, typically a government, which is extremely unlikely to default during the life of the bond.

Now we have our proxy, we can step back and remind ourselves that the return will consist of two elements. First, a return for waiting, referred to as the "real risk-free rate," and second, a return to compensate for the decline in the value of the currency relative to the energy spent, estimated from the expected rate of price inflation. Earlier we showed this relationship using the expression:

$$R_F = (1 + R_{Real}) \times (1 + \textit{Inflation}) - 1$$

Below, in Table 11.9, we present the data for US government bonds, since we are doing our forecast for GlobalTire's North American agricultural tire business.

As of May 2013, the risk-free rate varies from 0.04% for the 3-month horizon to 3.1% for the 30-year horizon. This difference reflects different inflationary expectations as well as different expectations of the real rate. In the 30 years hence, as inflation in the US is expected to be approximately 2.22%[4], the implied real risk-free rate is 0.88%.

But let's not jump too quickly, since we know that this seeming finding may very well be evidence of a red line metric misleading us. Why might the government bond rate be a poor indicator of the true risk-free rate? Because it is only the red line version of what we are looking for, and it may be impacted by supply-and-demand factors such as "flight to security" as multiple European governments are anticipated to default in the coming decade. This would result in the "pricing" of the US government bond not correctly reflecting the actual real risk-free rate. On that basis, we will consider instead the estimate of the risk-free rate provided by the return expected to be delivered by investments in the real economy, as discussed above. We have used the past data on this to estimate the real risk-free rate of 2%. We may be asking whether this rate applies to the future as well. According to the International Monetary Fund (IMF), the forecast growth in constant-price US GDP is between 2% and 3.5% through 2017[5]. Another estimate of the expected future growth in GDP is provided by the World Bank, which estimates growth in constant-price, US GDP of between 2.1% and 2.8% through 2014[6]. We

---

[4]There are various sources providing inflation forecasts, but we refer here to a February 2012 study by PriceWaterhouseCoopers (PwC) that provides inflation forecasts for many countries to 2021. The yearly forecast for the US starting in 2012 is 2.3%, 2.4%, 2.2%, 2.3%, 2.3%, and then is forecast to average 2.2% annually until 2021. We used the average of these.

[5]International Monetary Fund, World Economic Outlook Database, October 2012.

[6]World Bank, The global outlook in summary, June 2012.

Table 11.9: Data for US government bonds

| U.S. Government Treasury Securities (Treasury constant maturities, yields in percent per annum) | | | | | |
|---|---|---|---|---|---|
| | 2013 May 13 | 2013 May 14 | 2014 May 15 | 2014 May 16 | |
| **Nominal** | | | | | |
| 1-month | 0.02 | 0.01 | 0.01 | 0.00 | |
| 3-month | 0.05 | 0.05 | 0.04 | 0.03 | |
| 6-month | 0.08 | 0.09 | 0.09 | 0.08 | |
| 1-year | 0.13 | 0.12 | 0.12 | 0.12 | |
| 2-year | 0.24 | 0.26 | 0.26 | 0.23 | |
| 3-year | 0.40 | 0.41 | 0.40 | 0.37 | |
| 5-year | 0.83 | 0.85 | 0.84 | 0.79 | |
| 7-year | 1.30 | 1.33 | 1.32 | 1.25 | |
| 10-year | 1.92 | 1.96 | 1.94 | 1.87 | |
| 20-year | 2.73 | 2.77 | 2.76 | 2.69 | |
| 30-year | 3.13 | 3.17 | 3.16 | 3.09 | |
| **Inflation indexed** | | | | | |
| 5-year | −1.19 | −1.18 | −1.17 | −1.18 | |
| 7-year | −0.72 | −0.71 | −0.69 | −0.73 | |
| 10-year | −0.41 | −0.37 | −0.36 | −0.40 | |
| 20-year | 0.30 | 0.35 | 0.36 | 0.31 | |
| 30-year | 0.69 | 0.74 | 0.75 | 0.70 | |

Source: www.federalreserve.gov, May 22, 2013.

take the average of the forecast to obtain an average annual growth rate of 2.89%. When adjusted for expected inflation, we obtain an estimated risk-free rate of:

$$R_F = (1 + 2.89\%) \times (1 + 2.25\%) - 1 = 5.2\%$$

This number is substantially higher than the estimate from US government bonds, but at the same time it results in numbers that fit corporate bond pricing (where the "flight to safety" phenomenon should have less of an impact on market price). Furthermore, it fits the macroeconomic understanding of the risk-free rate as the combination of expected inflation and expected real productivity growth, and it fits the bond pricing data where the "flight to safety" is less likely to distort prices. For these reasons, we will use this number as our estimate of the risk-free rate.

The next step is to estimate the riskiness of the company itself. Remember from Chapter 3 the vital idea that the only risk we are interested in is the non-diversifiable risk that GlobalTire shares with Nature's portfolio, measured by GlobalTire's beta, $\beta_{GlobalTire}$.

As a first pass, we could look at the estimate of the beta provided by any of several data services, like Bloomberg, Google Finance, Yahoo Finance, or Thomson Worldscope. From Thomson Worldscope we obtained an estimate for GlobalTire's beta of 1.41. But recall from Chapter 4 the impact of leverage on the expected future cash flows, the riskiness of the firm (none), and the relative riskiness of the various claims on the firm, such as equity and debt. In particular, as the proportion of financing from debt increases, the relative riskiness, measured by beta, of both debt and equity will also increase to reflect the shift in the distribution of the firm's cash.

The underlying riskiness of the firm does not change when the proportions of debt and equity financing change. But it is because the underlying risk does not change that the *relative riskiness of each type of risk* must adjust, thereby keeping the resulting average equal to the unchanged underlying risk. We capture this with the expression:

$$\beta_{Global\ Tire} = \frac{D}{D+E}\beta_{Global\ Tire\ debt} + \frac{E}{D+E}\beta_{Global\ Tire\ equity}$$

Note that the estimate from Thomson Worldscope is for GlobalTire's publicly traded equity, $\beta_{GlobalTire\ equity}$, and not for the underlying company.

To obtain an estimate for the underlying company's beta, $\beta_{GlobalTire}$, we must "unlever" the equity beta to correct for the fact that the publicly traded equity is made riskier than the underlying company by the presence of debt, which has a priority claim to the company's cash flows.

We then obtain estimates for the value of GlobalTire's future cash flows to be distributed respectively to equity and debt, convert these to present value, and substitute them in for $E$ and $D$ in the above expression in order to solve for $\beta_{GlobalTire}$. To estimate the value of GlobalTire's equity, we will make use of the fact that GlobalTire is publicly traded. As of early 2013, GlobalTire's share price is about US$68, with 180.5 million shares outstanding, for an equity market capitalization of $12.29 billion.

From GlobalTire's balance sheet we find total interest-bearing debt of roughly $4.94 billion and cash of $2.47 billion, so that the *net* debt is approximately $2.47 billion. We note that GlobalTire's debt is rated BBB by Standard & Poor's, and BAA2 by Moody's. This would suggest a debt beta of approximately 0.10. As stated above, we obtained an estimate of GlobalTire's equity beta of 1.41. Inserting these estimates into our equation, we obtain an estimate for GlobalTire's company beta of 1.19:

$$\beta_{Global\ Tire} = 0.16 \times 0.10 + 0.84 \times 1.41$$

$$\beta_{Global\ Tire} = 1.19$$

But again we must be careful not to jump the gun. We cannot simply plug this number directly into the Capital Asset Pricing Model (CAPM) formula to obtain an estimate of GlobalTire's unlevered cost of capital, because we have additional data available that will improve our estimate of the OCC for the business opportunity at hand. Remember, what makes each company in a particular industry unique is, by definition, diversifiable risk from the fully diversified portfolio's perspective. The other companies in the tire industry should therefore have the same unlevered beta, and unlevered OCC as GlobalTire and the proposed investment. We can introduce a larger sample into our estimate of the unlevered OCC for the business opportunity by including the other companies in the industry:

Table 11.10: Estimating bets using comparable company data

**Estimating Beta**

| Company | Levered beta | Debt (000's) | Cash (000's) | Market equity (000's) | E/(D + E) | D/(D + E) | Debt rating | Debt beta | Unlevered beta |
|---|---|---|---|---|---|---|---|---|---|
| GlobalTire | 1.41 | 3,793 | 1,933 | 9,455 | 0.84 | 0.16 | BBB | 0.102 | 1.19 |
| Goodyear | 2.33 | 5,631 | 2,083 | 2,594 | 0.42 | 0.58 | BB | 0.132 | 1.06 |
| Bridgestone | 0.75 | 675,823 | 254,586 | 1,447,321 | 0.77 | 0.23 | A | 0.086 | 0.60 |
| Continental | 1.34 | 8,555 | 1,393 | 13,710 | 0.66 | 0.34 | BBB | 0.102 | 0.92 |
| Hankook | 0.78 | 2,797,481 | 938,063 | 7,152,926 | 0.79 | 0.21 | AA | 0.064 | 0.63 |
| Pirelli | 1.39 | 1,771 | 788 | 4,334 | 0.82 | 0.18 | BBB | 0.102 | 1.15 |
|  |  |  |  |  |  |  |  |  | 0.93 |

This larger sample size yields an estimated unlevered beta of 0.93. This is the figure we should insert into the equation for the CAPM. Our estimate of the risk-free rate for US dollars is 5.2%, and of the market risk premium, 4.5%. We can now estimate the opportunity cost for any business opportunity in the tire industry as:

$$OCC_{GlobalTire} = 5.2\% + (0.93 \times 4.5\%) = 9.37\%$$

Having arrived at an estimate for the unlevered OCC, we have a figure by which to discount our estimate of the expected future free cash flows. With both sides of the value equation now captured, we are almost ready to estimate the value of the business opportunity before us.

First we must factor in tax benefits from the interest payments. We estimate interest payments from the debt provided by the proposed investment; then we estimate the tax savings generated by these interest payments (known as interest tax shields), and discount these at the opportunity cost of debt for the forecast period (and the OCC for the firm for the forecast period, where debt is assumed to be maintained at a constant ratio of firm value on average). We could also use the WACC method which adjusts the discount rate (i.e., the OCC) to capture the tax benefit of the interest payments. While the two approaches are theoretically similar, they are not exactly the same and will yield different results in practice, as shown below:

## Table 11.11: Valuation using both APV and WACC methodology

| Valuation: APV & WACC | 2014 | 2015 | 2016 | 2017 | 2018 | 2019 | 2020 | 2021 | 2022 | 2023 | 2024 | 2025 |
|---|---|---|---|---|---|---|---|---|---|---|---|---|
| $ thousands | | | | | | | | | | | | |
| Free cash flows | (31,030) | (4,705) | (3,435) | (1,291) | 1,159 | 3,104 | 5,154 | 5,481 | 5,816 | 6,157 | 6,506 | |
| Interest tax shields | | 119 | 119 | 119 | 119 | 119 | 119 | 125 | 130 | 136 | 141 | |
| Final year growth rate (2022–2023) | 4.0% | | | | | | | | | | | |
| Growth rate after 2023 | 4.0% | | | | | | | | | | | |
| **APV Method** | | | | | | | | | | | | |
| Discount rate for free cash flows | 9.4% | | | | | | | | | | | |
| Unlevered OCC discount factor | 1.00 | 0.91 | 0.84 | 0.76 | 0.70 | 0.64 | 0.58 | 0.53 | 0.49 | 0.45 | 0.41 | 0.37 |
| Discount rate for interest tax shields | 5.7% | | | | | | | | | | | |
| Debt OCC discount factor | 1.00 | 0.95 | 0.90 | 0.85 | 0.80 | 0.76 | 0.72 | 0.68 | 0.64 | 0.61 | 0.58 | 0.55 |
| PV (FCF)@ unlevered OCC | (31,030) | (4,302) | (2,872) | (987) | 810 | 1,984 | 3,012 | 2,929 | 2,841 | 2,750 | 2,657 | 21,321 |
| PV(Int tax shields) @ debt OCC | – | 113 | 106 | 101 | 95 | 90 | 85 | 85 | 84 | 83 | 81 | 1,020 |
| **NPV (APV Method)** | **1,057** | | | | | | | | | | | |
| **WACC Method** | | | | | | | | | | | | |
| WACC | 9.0% | | | | | | | | | | | |
| WACC discount factor by year | 1.00 | 0.92 | 0.84 | 0.77 | 0.71 | 0.65 | 0.59 | 0.55 | 0.50 | 0.46 | 0.42 | 0.39 |
| PV (FCF)@ WACC | (31,030) | (4,315) | (2,889) | (995) | 820 | 2,014 | 3,066 | 2,990 | 2,910 | 2,825 | 2,737 | 22,825 |
| **NPV (WACC Method)** | **958** | | | | | | | | | | | |

The final step (really!) in the exercise is to try to understand, as well as possible, the economics of the business and its industry so that we can perform a common-sense review of the estimates we have derived. Throughout our calculations we have been as diligent and thorough as possible, taking into account every piece of data available to us, and using it as intelligently as possible. We will conclude the exercise with a data-driven sense-check of the numbers.

To obtain the estimates for each item, we made a conscious effort to combine both historical data and an analysis of the current business opportunity. We will resist examining the end result of our forecast and testing our raw numbers again. To be meaningful, we will compare the implied outcomes from our inputs with historical numbers for which we have a reasonable benchmark.

For example, an implied outcome from our forecast is the Earnings Before Interest and Tax (EBIT)/Revenues ratio. We also have EBIT/Revenues ratios from the past. We can test the sensibility of our estimates by comparing the implications of our forecast for the future EBIT/Revenues ratio against past data. In the recent past, GlobalTire has averaged EBIT/Revenues of between 5.4% and 9.9%. For our forecast the average is 3.2%, with lower numbers in the early years and increasing once the project is up and running to stabilize at 11.3%. This is clearly a little higher than the firm average, but is reasonable for a premium product.

Moving to another item, the recent Cost of Goods Sold (COGS)/Revenues ratio for GlobalTire has been between 65% and 68%. Our forecast yields an average of 66% – even though this is a premium product, the new launch is expected to be challenging, and the agricultural market is difficult, so this estimation is reasonable for this segment.

A third number we can calculate from recent data is GlobalTire's Return on Invested Capital (ROIC), which has averaged between 3.9% and 10.4% .Our forecast average is 7.0%, but it rises during the early years of the project to reach a high of 13.3%. The reason for this difference involves the nature of the project and the net fixed assets. Whereas Net Fixed Assets (NFA) for GlobalTire overall are continuously and smoothly replenished, for an individual project this is not the case. The NFA for the proposed tire initiative is much more "chunky" given the substantial ($30 million) upfront investment followed by modest maintenance and refurbishments in the years to follow. Such an investment pattern generates an estimate of invested capital higher than it would be for the company as a whole. Though the profit from the project is expected to rise over time, Accounting's estimate of the invested capital will go down.

We have carefully observed the mechanics of forecasting, discounting, and estimating the OCC for the proposed new investment. We have attempted to use all available data both to generate our estimates with minimum bias and also to check their implications using some important ratios. When all is said and done, the only thing certain is that our estimate of the value of the project is incorrect. There is likewise no doubt that each and every number we plugged in is incorrect.

But that is not the point of the Blue Line Imperative. The point is that we have held true to the data-driven, logic-based approach, and we can confidently say at this point that we have done all we can and there is limited opportunity for greater precision. The number we have generated indicates positive value for the project. We can continue to seek better data to challenge our numbers. We might try to improve our analysis or the quality of the logic we have applied. But in the end, if we choose to recommend that the North American agricultural tire business is value-creating and should be taken up by the company, we will certainly have done our homework.

# CHAPTER TWELVE

# BECOMING A BLUE LINE MANAGER

Though we've cited numerous examples throughout this book of the Blue Line Imperative, both positive and negative, there may be no better example of the blue line in action than professional basketball player Shane Battier.

A native of Birmingham, Michigan, Battier attended the perennial college basketball powerhouse, Duke University in North Carolina, before beginning his National Basketball Association (NBA) career in 2001 with the Memphis Grizzlies. After lasting five years in Memphis, Battier was then traded to the Houston Rockets.[1] While in Houston, he became the subject of a profile in *The New York Times Magazine* by financial writer Michael Lewis. In this article, *"The No-Stats All-Star,"* Lewis described Battier as a player whose conventional statistics – points scored, rebounds, assists, blocked shots, steals – hardly shine in comparison to the league's elite players,[2] yet whose team somehow displayed an uncanny knack for winning whenever he was on the court. The question Lewis sought to answer was how Battier could have such a positive influence on his team despite mediocre numbers and only marginal athleticism, admittedly as

---

[1] Battier was later traded back to the Grizzlies. In late 2011, he signed with the Miami Heat.

[2] All quotes in this section, unless otherwise noted, come from Lewis's article, The No-Stats Star, *The New York Times Magazine*, February 15 2009. See also http://www.nytimes.com/2009/02/15/magazine/15Battier-t.html

measured by the freakish standards of the NBA. As the Rockets' General Manager (GM) was quoted as saying, *"[Battier] can't dribble, he's slow and hasn't got much body control."* Such a statement – from Battier's own GM – begs the question how someone like that could not only survive in a competitive cauldron like the NBA, but have such a long, productive and successful career.

Lewis described Battier's game as *"a weird combination of obvious weaknesses and nearly invisible strengths. When he is on the court, his teammates get better, often a lot better, and his opponents get worse – often a lot worse."* Despite mediocre personal stats, Lewis observed, Battier is that rare player who, it would seem, really does elevate the performance of those around him.

More specifically, he elevates their statistics due to his unselfish, value-oriented behavior on the court. If Battier is not in the best position to snatch a rebound, for example, he will routinely tap the ball to a teammate who is. The teammate gets statistical credit for the rebound, even though it would not have happened without Battier's blue line act. Fans of the NBA may argue that lots of players do this sporadically, but what is unique about Battier is the consistency with which he demonstrates these purely value-driven acts. His game is full of such examples, a myriad of subtle things he does that a casual observer (and sometimes even a seasoned one) wouldn't notice, but each of which makes a marginally positive contribution to his team. Aggregated over the course of a game, these actions often contribute enormously to the difference between Battier's teams winning or losing.

What makes Battier doubly fascinating is the fact that basketball is arguably the one major team sport beset by agency problems – the problems that arise in any organization where there is a true separation of ownership and management, since managers or employees will advance their personal goals while destroying value for the company's owners. In a sports context, agency costs are seen whenever a player does something that he perceives to be in his personal interest even if it hurts his team. In most team sports, agency conflicts are minor. A baseball pitcher who tries to achieve a low Earned Run Average (that is, gives up the fewest runs possible to the opposition) will virtually always help his team by doing so. A batsman in cricket concerned only with scoring as many runs as possible is at the same time increasing his team's chances of winning.[3]

Basketball is different, insofar as its players may perform any number of individually valuable acts that nonetheless do not contribute to team success, and in fact often detract from it. Worse, professional basketball players often have powerful incentives, as part of multi-million-dollar contracts, to make certain indicators look better regardless of the effect they may have on team value. In basketball, there are any number of things a player can do to make his own stats look better while hurting his team. As Lewis writes, *"The game continually tempts people who play it, to do things that are not in the interest of the group."* For example:

> "A point guard might selfishly give up an open shot for an assist. You can see it happen every night, when he's racing down court for an open layup, and instead of taking it, he passes it back to a trailing teammate. The teammate usually finished with some sensational dunk, but the likelihood of scoring nevertheless declined."

Essentially, the point guard – the player on a basketball team who usually brings the ball up the floor and "directs traffic" – passes up a likely two points to make a pass that allows his teammate to score and get credit for the two points instead. On the surface, and to the majority

---

[3] However, agency costs do occasionally crop up in cricket. Batsmen have been known to hurt their teams in pursuit of personal milestones, such as the much-coveted "century," in which an individual batsman scores 100 runs or more in a single innings. In a 2011 test match between Australia and Sri Lanka, one of Sri Lanka's batsmen was on the verge of scoring his first century in test cricket. The closer he got, the more nervous he became. The free-flowing strokes that got him past 80 disappeared, as he seemed more interested in not getting out than in scoring runs. The problem for his team is that Sri Lanka were on top, and their only hope of winning was to press on quickly, then put Australia in to bat and try to bowl them out. But the batsman took too much time to score his runs. He reached the milestone, but in so doing damaged his team's chances of winning the match. In the end, Australia easily secured a draw.

of fans, it seems like a selfless act. But if we analyze the act from a blue line perspective, it reveals itself as value-destroying, since, the talent level of NBA players aside, the act of making an extra pass increases the likelihood of something potentially going wrong. The guard inadvertently makes a bad pass. The teammate doesn't field the pass cleanly. The other team has more time to get defenders in position. And so on.

In other words, the action reduces the expected value of the point guard's behavior. So why do it? Because to a point guard, typically judged more on his ability to pass the ball and help his teammates score rather than on scoring himself, any assist is worth more than two points. If a player at this position accumulates a lot of points but fewer assists than his peers, he may no longer be viewed as a "pure" point guard, which means less money with his next contract. He is not judging expected value according to the number of points his team scores, which is the only true predictor (along with how many it allows the other team to score) of whether it will win or lose. He is judging it based on assists, which are more valuable to him.

A similar example can be found in the work of economist John Huizinga and sports statistician Sandy Weil, who focused on blocked shots (when a defender blocks an opposing player's attempt to score) as an indicator.[4] The universally acknowledged master of this art is the Orlando Magic's Dwight Howard, perennially one of the standout "big men" in the NBA.[5]

Huizinga and Weil wondered whether there was a more nuanced way of looking at this statistic. They noticed that, while Howard often leads the league in blocks, an inordinate number of the shots he swats away tend to wind up in the stands because of the force with which he performs the act. In other words, while there is no arguing that Howard's blocks look customarily spectacular and often whip the crowd into a frenzy, the opposing team usually retains possession, that is, they get the ball right back.

---

[4]The Huizinga and Weil study is described in Tobias J. Moskowitz and L. Jon Wertheim, *Scorecasting: The Hidden Influences Behind How Sports are Played and Games are Won,* (New York: Random House, 2011), pp. 86–91.
[5]As of this writing (2013), Howard had just been traded to the Los Angeles Lakers.

Blocked shots by another of the game's stellar proponents of the act, the San Antonio Spurs' Tim Duncan, end up in the hands of teammates more often than do those by Howard – *much* more often. For example, in 2008 the average Tim Duncan block contributed 1.12 points to San Antonio, while the average Dwight Howard block contributed only 0.53 points to Orlando. Though Howard comfortably led the league in the *statistics*, the *value* of his blocks was considerably less than Duncan's.

In their discussion of this research, University of Chicago professor Tobias Moskowitz and sports journalist Jon Wertheim ask, *"If the value of an act is what matters rather than the act itself, then why do we – and the NBA – count blocks rather than value them? The short answer is probably the correct one: counting is easy, measuring is hard. You don't have to look far to see this dichotomy at play in any number of facets of life and business. It is much more common for people to count quantities than to measure importance, and as a result to make misguided decisions."*[6] The red line seduces us because it is so easy to see. Managing the blue line, on the other hand, is a more elusive effort.

Which brings us back to Shane Battier. After an exhaustive study of his player, the Rocket's general manager changed his tune, asserting that Battier is *"the most abnormally unselfish basketball player"* he has ever encountered. He helps the team, the GM now says, *"in all sorts of subtle, hard-to-measure ways that appear to violate his own personal interests."* The GM has it right this time. Battier is a pure blue line phenomenon, focusing all of his talents and energy on whatever it takes to help his team win even if at the cost of his own statistics.

It is interesting to note that because Battier's value-creating actions are at times so subtle, when the opposing team's star scores fewer points than usual, a common occurrence when Battier is the defender, Battier rarely gets the credit. Instead the star will usually explain his lackluster performance by saying something to the press like *"I had a bad night"* or *"The shots just weren't falling."* Never will you hear the star opine, *"Shane must have really done his homework, because he knew exactly how to shut me down tonight."*

Among the players Battier has been tasked with guarding is the Los Angeles Lakers' superstar Kobe Bryant. Because Battier is so data-driven,

---

[6]Ibid., pp. 90–91.

his team supplies him with a mountain of information before facing the Lakers, something they don't normally do for other players. Why do they expend the effort and resources to do this? Because they know that, like any good blue line manager, Battier will study the data in order to unearth tendencies or weaknesses he can exploit. As one might expect from an elite player, Bryant has no glaring weaknesses, so the exercise is not an easy one. But true value-orientation demands examining as much data as is available to try to detect insights, and Battier does exactly this.

Battier further demonstrates a blue line approach by focusing on the process and not the outcome. He attempts to lure Bryant into his "zones of lowest efficiency," meaning those parts of the court or types of plays where his chances of scoring are, according to the data, lower than usual. Battier understands fully that even the best strategy can never be fool-proof, and that we live in a world of probabilities, not guarantees, so he does not expect to shut Bryant down completely. Such was the case during a recent game Lewis describes. With just a few seconds left in the contest, Houston took a one-point lead. The Lakers had the ball and enough time left for one final shot. Naturally the ball was fed into Bryant's hands. Anticipating this, Battier, covering Bryant, forced him to a spot far from the basket, in fact well beyond the 3-point arc, a particular spot on the floor from which Bryant misses more than 80% of the time. With the clock running out, Bryant was forced to launch the ball. Lewis describes what happened next:

> "Battier looked back to see the ball drop through the basket
> and hit the floor. In that brief moment he was a picture of
> detachment, less a party to a traffic accident than a curious
> passer-by. And then he laughed. The process had gone just as
> he hoped. The outcome he never could control."

Remember Pierre and Annette? Pierre invested in a positive-NPV (Net Present Value) project, but when the cosmic dice were rolled, he lost, at least in the particular instance we described. That we live among probabilities and not certainties means even value-oriented guesses are sometimes going to yield negative results in isolation. But that doesn't change the fact that you always want a guy like Battier on your basketball team, just as you always want a guy like Pierre to work for your company. With enough people believing in and pursuing the Blue Line Imperative, you will eventually come out on top.

## How can I Become More of a Blue Line Manager?

One of the more striking aspects of Shane Battier's story is that no one told him to behave the way he does. Though he has had the good fortune to be coached by some of the most respected minds in the game, many others have received similar tutelage. That Battier's actions are so atypical demonstrates that his blue line behavior is largely self-taught.

Seminar participants often tell us that while they understand and personally believe in the Blue Line Imperative, applying the concepts within their organization is a different thing altogether. The most common reason for this is feeling stymied by their own top management, and the biggest single impediment, the pervasiveness of indicator-driven bonus plans.

Changing such a compensation culture is hard and, at least in the short run, practically impossible. This we understand. But most importantly, change happens gradually, bit by bit, and there are distinct steps you can take to become more blue line driven. Like Shane Battier, you don't need anyone's permission to do them, and eventually someone is bound to notice the value such behaviors create, hopefully creating a ripple effect of blue line actions. You may not be profiled by Michael Lewis in *The New York Times Magazine*, but that's not your goal, just as it isn't Battier's. Your goal, like his, is to create value, and therefore to increase your team's chances of success.

Following are some things you can do right now to make yourself, and by implication your organization, more focused on the blue line.

## Design Your Own Blue Line Questions

As we have said repeatedly, there are no definitive answers in business, only questions. Creating value therefore comes down to asking the right questions. By asking the right questions, you gather the right data. By gathering the right data, you create the possibility of gaining meaningful insights into the real value drivers. By gaining such insights, you enable real learning.

As you proactively design your blue line questions, remember above all not to do anything that might compromise the integrity of the data. Should the data be manipulated, genuine learning cannot occur. The data must be allowed to speak for itself; there must be no attempt to steer what it has to say.

If we adopt a red line approach, an entirely different set of questions is posed, leading the firm and its people to focus on indicator outcomes, and not the underlying drivers of value. To manage on the basis of value alone, you must begin by asking a series of blue line questions. Their purpose is, of course, to yield a set of responses that define the current business model – *what* your company does to create value, *who* its customers are, and *how* it transforms energy into products and services customers are willing to pay for. You must ask blue line questions regarding each of these three components. Through them, your goal is to clarify first what the business strives for and the resources it uses to try to achieve this goal, and second, what behaviors, actions, and decisions in the organization can be deemed value-creating. In other words, which behaviors raise the blue line, and which do not?

In the appendix, we offer a set of blue line questions applicable to almost any enterprise. The idea is to adapt the list to your own situation, and to see how many more you can think of. They are diagnostic questions, and they can be, and should be, as detailed and operational as you need. You also ought to include questions that are more industry- or sector-specific.

Not only do we encourage all enterprises to develop a comprehensive list of blue line questions, but top managers should ask their reports to do likewise for their own units or departments. Show them the questions you use, urge them to put them into practice as well, and encourage them to develop their own. The more questions designed around value are used at different levels throughout the organization, the more likely the organization as a whole will succeed.

### Identify the Key Value Drivers for Your Business Unit or Department

The blue line questions are designed to help you uncover the true drivers of value in your area. Do everything you can to make sure you really are identifying the *drivers of value* and not simply listing the activities your department performs. Remember critically that targets are not value drivers. Think back to the sales calls example discussed in Chapter 5. Simply setting a target of a certain number of calls per hour only yields numbers of calls, not the *impact* those calls have on your business. The blue line questions you and your colleagues have compiled should help. If they don't, well, you haven't been asking yourself the right questions.

It is these value drivers, not indicators, that must be emphasized in all of your communications with direct reports. Eschew the indicator-driven mentality that holds out targets as the be-all and end-all of operations regardless of their actual value contribution. To every extent possible, espouse and encourage a value mindset, always urging your colleagues to question what they do from a blue line perspective. You are not suggesting ignoring the indicators. You are trying to convince others to use them properly, as truthful reflections of how well you are managing the business.

One very important outcome of talking about value drivers instead of indicators, is avoiding the reflex of finger-pointing when a target isn't reached. People should have the courage to fail. They can't be expected to possess such courage if you don't establish the conditions in which failures are defined as opportunities for learning. Inevitably, things will arise that resemble targets. Before deciding what they really are, consider them closely. A value-driven focus will help you examine them from the right perspective, which is how they relate to performance.

### Always Promote Honesty in the Measurement of Key Performance Indicators

Communicate unequivocally to your people that the primary role of indicators is to help all of you to learn, and never to encourage judgment or blame. At the risk of sounding repetitive, make the point clearly that because KPIs are typically used to define targets, they often become a source of blame, which is obviously poisonous for any business. Tell those within your charge that they are required to avoid this so that as a team you can use KPIs for their proper purpose – to learn about what drives value in your business.

## If You're Senior Enough, Try to Get Rid of Key Performance Indicator-Driven Pay

We get asked the following questions all the time: *"If we shouldn't pay people based on KPI targets, then how should we pay them? How can we properly motivate them if we don't have financial incentives linked to measurable performance?"*

Seldom have we run a seminar in which at least one person hasn't asked these questions. One of the problems implicit in them is that, if we pay people to do a certain something, can we be sure that the thing they are doing is creating value for the company? The real question to ask is what your people are truly being paid for – better management of key value drivers or the outcome for a specific indicator? We have argued repeatedly that if it is the latter, there is a very small chance you are focusing your people on value creation. Their energies instead are directed toward meeting or exceeding indicator targets and of course these targets are never more than, at best, imprecise measures of what you want your people to do.

When we discuss with our participants the various things their people consciously or unconsciously do to reach KPI targets, inevitably someone will suggest that this problem can be nipped in the bud by holding managers accountable for multiple indicators. The idea is that people can game one, or maybe two, indicators, but they cannot game all of them.

This practice may indeed help in part, but no incentive system can be sufficiently "complete" to capture every aspect of a person's job. As economist John Roberts explains, *"Contracts cannot fully specify the desired behavior in every circumstance and so cannot fully resolve motivation problems."*[7] A "crowding out" phenomenon occurs when managers direct their attention to behaviors captured in the performance measurement system while at the same time ignoring those that are not.

The problem becomes more acute as one moves up the organizational ladder. Whenever managers are faced with tradeoffs in allocating resources – a challenge senior executives constantly face – their decision will be skewed in favor of those alternatives that offer the best prospect of deliv-

---

[7] John Roberts, *The Modern Firm: Organization Design for Performance and Growth*, (Oxford: Oxford University Press, 2004), p. 125.

ering on the KPI targets, whether or not those decisions are value-destroying. If your organization relies on indicator outcomes as the principal tools for determining bonuses and other rewards, this problem cannot be avoided. Try to imagine for a moment the full range of issues and responsibilities involved in a typical senior management position. How could one possibly create a set of indicators that captures all such relevant factors? The inevitable result is "incomplete" contracting, in which important things get left out.

A scene from the US television comedy The Big Bang Theory illustrates this point nicely. The scene opens with Leonard, a Caltech physicist, telling his girlfriend he has been invited to Switzerland to visit the CERN Super Collider, and that he plans to take her with him so that they can do a bit of skiing and celebrate Valentine's Day together. But Leonard's asocial roommate, Sheldon, also a Caltech physicist, has different ideas. He believes he should be the one who gets to accompany Leonard on the trip. Sheldon produces a binder containing their "Roommate Agreement," a document that spells out in excruciating detail how the two friends are to act in nearly every contingency. Keen to see the Super Collider, Sheldon invokes the "friendship rider" in "Appendix C, future commitment #37," which stipulates that if either friend is invited to visit the Super Collider, he shall bring the other. (The contract also details what should happen if either roommate acquires superpowers or is bitten by a zombie.)

In this case, a contract was created that did indeed cover nearly every circumstance, leading to a virtually complete contract.[8] The very existence of such a document is played to great comic effect, but besides its being part of a fictional television show, it raises a troubling question: to what extent does Sheldon trust Leonard? The fact that he feels a contract is necessary to cover every contingency suggests he doesn't, though he would almost certainly say that he and Leonard are close friends.

This is precisely the issue that arises when KPIs are piled one on top of another in the hope of covering the full range of an individual's responsibilities, or to make gaming of a particular indicator more difficult

---

[8] Very close, yes, but complete, no. When Leonard's girlfriend asks if the contract says anything about what happens if one of them gets a girlfriend, Sheldon responds, "No, that seems a little farfetched."

by adding on other indicators that measure similar behaviors in a differ-ent way. The ever-present danger is that detailed contracting in the form of explicit incentives tied to KPIs will often be interpreted – justifiably – as a signal of distrust. In trust-based organizations, it is expected that people will rely upon their own discretion and goodwill to do what is right for the company without having detailed contracts or indicator targets that specify in detail what they are expected to do. Trust is the default mechanism that fills in the gaps necessarily arising from incom-plete contracting.[9]

We believe just about everyone has the built-in ability to distinguish value creation from value destruction. As we have discussed earlier, if we didn't possess this ability, it's hard to imagine how we could have ever advanced beyond our caveman origins. Certainly this trait is more finely honed in some of us than in others, but we all have it to some degree. One reason we are convinced of this comes from the many admissions of our seminar participants and corporate clients that they have fre-quently taken actions contrary to the best interests of their company in order to reach KPI targets. More to the point, they feel bad about it, reinforcing the simple but crucial theme that value creation makes us happy, while value destruction makes us sad. Each of us knows value creation when we see it, and each of us experiences a similar sort of feeling when we observe the opposite.

Make the objectives of the company plain to your employees, and reinforce them continuously. If you can get those around you to buy in

---

[9]An interesting example of near-complete contracting can be seen in the project-financing structures commonly used for large infrastructure and industrial projects, like, power plants, toll roads, port facilities, theme parks, or airports. A special-purpose entity is created for each project, thereby shielding other assets owned by a project sponsor from the consequences of a project failure. In the largest of such projects, the contracting agents will number in the thou-sands, once all of the project's suppliers are considered. Because so many players are involved and various forms of financing used, the nexus of contracts that defines the special-purpose entity can be extraordinarily complex. Provisions will specify, in great detail, the priority of claim and expected timing for each cash flow, what happens in the event of cash shortfalls, and how any positive residual cash flows will be allocated. These arrangements are about as close to "complete contracting" as you will find.

to these objectives and to your business model, they can be trusted to figure out what they need to do to create value for the company. Gathering honest and reliable data, especially KPIs, will help guide them in their efforts, but it is their inherent sense of value creation that, if the right conditions are in place, will lead them to take actions that are good for the company. Follow the Google example. Establish a blue line environment, let people set their own value-based targets, and watch the company move forward.

### Set the Example

It isn't fair to expect those who report to you to be value driven if you yourself are a slave to the red line. For example, if you are a senior manager and you engage in earnings management (that is, accounting manipulation), the signal you are sending to your people is not exactly one of value orientation. Ask yourself whether there are any aspects of your own conduct that might lead others to question your integrity. Examine whether the standards you've set for yourself demand probity in the gathering and analysis of data. Be clear about whether you use indicators to unlock the true drivers of value. Be able to convince yourself that you are upholding The Blue Line Imperative. Only then can you ask others to follow suit.

### Develop Templates and Spreadsheet Skills that Allow You to Assess the Value Impact of Proposed Actions

Use NPV to the extent that it is practical to do so. Remember that understanding the cash flow impact of proposed decisions goes to the very heart of what it means to be value driven. For those operating decisions that don't require sizable capital investments, formal NPV models will not be useful. Instead, use spreadsheets, customized software, or any other possible tools that can help your people assess the value contribution of their decisions.

### Sketch Out Your Value Chain, Reaching Downstream to Your Customers and Upstream to Your Suppliers

Distinguish between your role in the chain and the roles played by others. How do you add value to the inputs received from suppliers, whether internal or external? Where are you most vulnerable to rivals?

As a second step, list the key activities of your business unit and determine which of them are value-creating, as defined by your unit being better at them than other businesses. If these activities are done better elsewhere, why haven't they been outsourced? And do you possess the capabilities to increase efficiency enough to justify these activities continuing to be performed in-house?

## Describe How Success is Measured in the Business – Or at Least in the Area You're Responsible For

To what extent is there conflict between the way you measure success in your area and genuine value creation? Determine any dysfunctional behaviors that may arise because of the ways in which success is defined. Understanding these problems can help you recognize and eradicate value-destroying behavior at its source.

## Conduct a Value Audit

Much of what we've written in the previous pages has been devoted to describing what it means to be value driven. Below we attempt to encapsulate this thinking in a summary of blue line attributes you should hope to see in your business.

We would go even further in fact, and assert that these are attributes we would hope to see in any organization, even government and non-profits. The more organizations performing honest, no-holds-barred assessments of where they truly stand on value, the more collective value is created.

It is possible that a few of these attributes won't apply to you (for example, if your company is not publicly traded, you don't need to worry about share price), but most will. We have organized these attributes into six broad categories, and some will overlap. We urge you to look closely at these lists. For each attribute that exists in your area or organization, give yourself a pat on the back. For each one that doesn't, think about how to introduce it.

1. Decision-making and resource allocation.
   - When making investments, value creation is the only criterion that matters to us.

- We understand the difference between value and price, and we focus all our efforts and attention on the former.
- We don't try to "manage" share price.
- We don't justify decisions on the basis of expected share price impact.
- We allocate capital based on NPV.
- We understand that value is based on *the* expected cash flows, not the personal opinions or beliefs of our managers.
- We make capital structure decisions (debt vs. equity) on value-based criteria only.
- We understand and agree that the only growth worth pursuing is positive-NPV growth.
- We acknowledge that outperforming rivals is necessary but not sufficient for value creation (investments must also be positive-NPV).
- We know how much we can grow without destroying value.
- We don't allow sunk costs to influence decision-making.
- We don't have a "grow-at-all-costs" mentality.

2. The Opportunity Cost of Capital.
   - We use the OCC as our investment benchmark, not the cost of financing.
   - When valuing capital investments, we ignore financing.
   - We don't allow diversifiable risk to influence the discount rates we use for valuing capital investments.
   - In valuing emerging-market investments, we incorporate incremental risk in the expected cash flows, not in the discount rate.
   - We use a rigorous statistical approach to estimate our OCC.

3. Fairness and trust.
   - Fairness is a pervasive attribute of our organization.
   - Fair process is present in all of our management systems.
   - Our organization is trust-based.
   - We have a team-based mentality.
   - We are open and transparent in our dealings with one another.
   - We don't have a steeply hierarchical business.
   - Employees at all levels have a genuine sense of pride in our organization.
   - Communication between bosses and subordinates is open, honest, and two-directional.

- We can tell the difference between good ideas that don't work out (Pierre) and bad ideas that do (Annette).

4. Learning.
   - Our organization is data-driven.
   - Our organization is learning-based.
   - We encourage experimentation and tolerate failure as long as both drive learning.
   - We don't let assumptions, opinions, or beliefs get in the way of organizational learning.
   - We systematically document learning and share best practices.
   - Dissent from lower-level staff is tolerated and encouraged.

5. Performance measurement and control.
   - We make a conscientious and continuous effort to identify value drivers.
   - We don't confuse KPIs with value drivers.
   - Target-setting is done honestly, without attempts to sandbag or build in deliberate buffers.
   - Our performance measurement system encourages truth-telling.
   - Signals flow freely from our customers without delay or distortion.
   - When an employee fails to achieve a KPI target, we discourage blame.
   - Our financial reporting function is not red line driven.
   - Our accounting numbers tell the truth.

6. Incentives.
   - Our people are rewarded based on their contribution to value.
   - We don't tie bonuses to the achievement of KPI targets.
   - We don't tie senior management bonuses to market share or EPS.
   - We acknowledge the role that probabilities and randomness can play in success, and we allocate recognition and rewards accordingly.
   - Our top managers do not have incentives to invest in negative-NPV growth.
   - Our appraisal and reward system is transparent and well understood.

Once you have reviewed these attributes, consider the value diagram in Figure 12.1 below:

## Figure 12.1: A generic value diagram

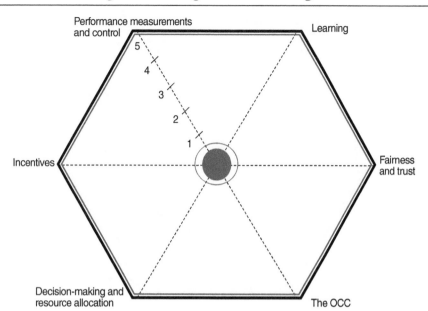

Each point of, the hexagon corresponds to one of the six attribute categories. The center of the diagram represents a pure red line perspective; the outer edges, a pure blue line perspective. Using your general assessment of the attributes, plot where you think your company or unit belongs on each of the six dimensions on a scale of 0 to 5, 0 being pure red line, 5 being pure blue line. Here's an example in Figure 12.2:

Figure 12.2: Value gap analysis for a hypothetical company

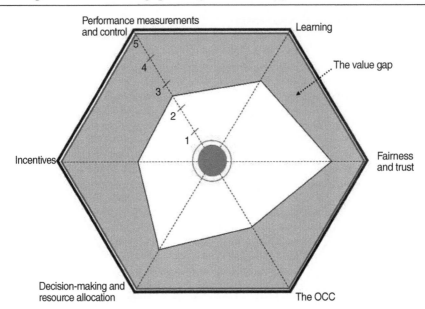

In the above case, the fictional business in question scores reasonably well on Decision-Making and Resource Allocation, Fairness and Trust, and Learning, but less well on the other three dimensions. Even for those dimensions in which the business performs strongly, there is still room for improvement.

Once the scores are determined for each of the six major categories and the dots are connected, shade the area between the curve and the outer edges of the diagram. The shaded area represents your value gap – the areas in which there is still room to achieve a pure value-driven business model.

## A Final Word

Value creation is hard. It's always been hard, and it isn't getting any easier. Markets are more competitive than ever, and despite recent pro-tectionist rumblings in some countries, are likely to stay that way. The reason for this is simple. As human beings, we wouldn't have it any other way. Competitive pressure in business is a necessary and unavoidable by-product of our collective desire to be happy. The Blue Line Imperative is a system by which we can gain sustainable competitive advantage by satisfying this desire.

We hope you have agreed with many of the lessons in this book and wish you luck in applying them in your business. May you benefit from the value-driven focus of Pierre combined with the good fortune of Annette. We would be grateful if you would share with us your experiences in putting these ideas into practice, and we are doubly grateful for any suggestions or comments you might have. Our undeviating goal is both to identify impediments to value creation and continue devising ways to overcome them. We enthusiastically invite you to join us in this endeavor.

kevin.kaiser@insead.edu
david.young@insead.edu

# APPENDIX

# BLUE LINE QUESTIONS

Here we present a series of questions designed to uncover insights into the true value drivers of your business – blue line questions, in other words. By asking these questions of yourself and those around you on a regular basis, you will help establish and maintain a culture oriented relentlessly toward value.

### What?

The focus here is on the company's **Value Proposition**, a statement on the products and services it sells to create happiness for a specific group of customers. Because businesses often serve several types of clientele, firms can have multiple value propositions. Think of your value proposition as the reason why customers turn to you instead of your competitors. The questions include:

- Do we have a competitive advantage? What are the main sources of that advantage? How sustainable is it? How will it be lost? To whom will it be lost?
- If we don't have a competitive advantage in our industry, who does?
- How has our corporate strategy changed in recent years? Why has it changed?
- What do we do differently from our rivals?
- Who are our key competitors, and which of them poses the most significant threat? Why? What are we doing about it?
- Why do our customers buy from us and not from competitors?
- Why do other customers buy from the competition and not from us?

- What features and attributes do our competitors offer that we don't?
- How do we help customers assess our value proposition?
- Which industry player is the leader in producing innovations that customers care about most?
- On what dimensions are we beating competitors? On what dimensions are we behind?
- Who are our competitors-in-waiting?
- How well do we compete on price? Quality? Service? Branding? Other attributes?
- What is the nature of the competitive landscape in our industry? How quickly is it changing? Are the changes driven by technology? Consumer preferences? Regulatory changes? Other factors? What are we doing to preempt these changes or ensure maximum reactivity to unanticipated changes in these factors?

## For Whom?

Here you are concentrating on your **Customers** – the only people who will truly decide whether what your company offers has value.

- What do we sell?
- To whom do we sell it?
- Why do they buy it?
- Do we go after the mass market or a niche market?
- According to what factors do we segment our customers? Income? Geography? Age? Gender? Others?
- What features and attributes do our customers value most, and how much are they willing to pay for them?
- How quickly do our customers pay?
- How common are customer complaints, and what do they tend to complain about?
- How quickly are complaints resolved? Do senior executives ever get involved?
- How do we raise awareness of our products and services among customers?
- How successful are we at retaining customers?

Questions regarding your company's **Revenue Model** are concerned with how you generate cash from each customer segment.

- Do revenues for each segment come from one-off sales, from recurring transactions, or from some combination of the two? What's the relative proportion?
- Are the prices we charge in each customer segment fixed or negotiable?
- Is revenue generated from sales of a physical product? From leasing? From fees? From other sources? From a combination? What are the different revenue streams, and how much does each contribute to overall sales?
- Are revenues always realized in the form of cash, or are other considerations sometimes given?

*How?*

Questions about the **Channels** through which your company reaches its customers.

- Do we rely on our own sales force, do we go through third parties such as wholesalers, or do we use some combination of the two?
- How exactly do customers purchase specific products and services?
- Are our processes and procedures for delivering products and services well documented?
- Are our processes for selling and delivering products efficient? How do we know?
- To what degree do we rely on web sales, if at all?
- How far down the distribution channel do we operate?
- If we aren't in direct contact with customers, who is? How do we ensure awareness of, and responsiveness to, customer desires if we aren't in direct contact?
- How do we market our products and/or services to customers?
- How effective is our marketing? How do we test this effectiveness?
- How is the sales force organized? By geography? Product line? Customer type? Some other way?
- Through what incentives do we motivate the sales force? Do we believe these incentives are effective?

Questions about the four types of **Resources** your company uses –
people, intellectual capital, physical capital, and finance – delve into the
sources of energy the business depends on to deliver its value proposition to customers.

*People*

- How many people do we employ?
- Through what incentives do we motivate our people to do well? Are
  there any counterproductive behaviors encouraged by our pay policies?
- Which performance indicators are linked to senior management pay?
- Do we have a stock option program? How far down the corporate
  hierarchy does it go? How is the size and value of option grants determined? What are the vesting rules, and do they encourage long-term,
  value-oriented thinking?
- Do we have an employee stock-ownership program? Does it cover all
  employees?
- How much division of labor is there within the company? Does the
  average employee perform a narrow or broad range of tasks?
- How many people per function do we have compared to the competition? Can we explain the differences?
- Are we effective at maintaining high morale and enthusiasm? What
  steps do we take to achieve this on an ongoing basis?
- How do we ensure that employees are receiving "value" from their
  jobs such as intellectual stimulation, autonomy, development, training,
  recognition for performance, a sense of acknowledgment, and respect
  for their contribution and effort?
- How steep is the organizational chart? How many steps exist between
  the Chief Executive Officer and the lowest-level employees? How many
  direct reports does the CEO have?[1] What are our key values and beliefs?
  Are they documented and disseminated? How?
- What constitutes acceptable and unacceptable behavior in our organization? Are these behavioral norms well understood?
- To what extent do we help our employees get the skills, knowledge,
  and tools they need to perform better at their jobs?

---

[1] Note that, in general, the greater the number of direct reports to the CEO, the
flatter the organization.

- Do the senior HR staff know the different headcounts at which various pieces of regulation kick in? Are systems in place to ensure compliance with health, safety, and employment laws?
- How often are performance reviews conducted? What criteria are evaluated and discussed in these reviews? What is the link, if any, between these reviews and compensation?
- To what extent are our employees computer-literate?
- Do we maintain a good safety record? Is there a process in place to ensure that accidents are reported swiftly to top management? What procedures are documented and enforced to ensure that processes and procedures are updated and improved to reduce accidents?

## Intellectual Capital

- What are our key intangible assets – brands, patents, copyrights, trademarks, licenses, human capital ("know-how"), customer relationships, supplier relationships, employee relationships, government relationships, etc?
- How much do we spend on intangibles each year? What measures do we use to track how effectively we manage these assets?
- Do we know what we want our website to accomplish? How well do we feel it satisfies this aim?
- How much do we rely on customized software? Do we maintain intellectual rights over it? Is the documentation written down and up to date?
- Do we have a logo or wordmark? Is it distinct and easy to recognize? Do we measure how much our logo and other branding resonates with customers?
- How much do we invest in R&D each year? How effective is our research program in terms of product or process innovation?
- How do we encourage innovation from our employees? Do we know if we're good at it? How do we measure this?
- Do we tend to lead or lag within the industry in terms of innovation?

## Physical Capital

- What fixed assets do we have?
- How quickly do we turn over inventory? How many days of purchases do we have in raw materials inventory? How many days for work-in-process and finished goods?

- Are our facilities clean, well-lit, and secure?
- Do those in Operations understand inventory management concepts such as "safety stock" and "economic order quantity"?
- How common are back orders? How quickly are they cleared?
- Do we provide our customers effective IT support? Do we have a Help Desk? Do we have a disaster recovery plan? Do we have offsite or "cloud" storage of critical files?
- Do we have significant underutilized capacity? If so, do we have a plan for using it?
- What measures do we use to track capital efficiency? How are we helping managers to understand and deliver on capital-efficiency improvements?

## Finance

- How much cash are we sitting on?
- Do we have a specific cash-management policy?
- What is our capital structure? Are debt levels high or low relative to competitors?
- Does our capital structure allow us to respond decisively to pricing threats from competitors?
- Can we raise large amounts of financing if needed to fund significant value-creating capital investments?
- Do we rely extensively on leasing? If so, why?
- What sort of risks do we face (e.g., foreign exchange, commodity price, interest rate, technological, political)?
- Do we have a hedging program in place? If not, why not? What instruments do we use for hedging?
- Is there evidence that corporate treasury uses derivatives for purposes other than hedging?
- Are our finance professionals able to explain their use of derivatives and other financial instruments clearly and convincingly?
- How do we assess capital-investment opportunities? What measures are used? NPV? Internal rate of return? Payback periods?
- Are our financial reporting policies clear and appropriate? Do they differ in significant ways from those of our rivals? If so, why?
- Is there evidence of any earnings management or unscrupulous accounting practices?

- Have our financial statements received an unqualified audit opinion from a reputable accounting firm? Is there any evidence of "opinion-shopping"?[2]
- Do we produce financial statements in a timely fashion?

Questions about your **Collaborators** describe the relationships you maintain with suppliers and other partners. Companies may also form joint ventures to reduce the risks associated with developing new businesses, or may form strategic partnerships with competitors who share a common interest (e.g., industry lobbying or agreeing on technical standards).

- What do we define as our "core" processes?
- For non-core processes, which are done in-house and which are outsourced?
- Why are the non-core processes done in-house *not* outsourced?
- Who are our suppliers?
- Which key suppliers are domiciled domestically, and which in other countries?
- How efficiently do we manage our supply chain? What benchmarks do we use? What kind of terms and working arrangements have we negotiated with suppliers?
- Do we have a just-in-time inventory system in any part of the business? If we don't, is our supply chain sufficiently responsive and efficient to support one?
- Do we have emergency plans in place in the event of supply-chain disruptions?
- Do we maintain any collaborative arrangements with competitors? What purposes do these arrangements serve? How do we determine their effectiveness?

Questions regarding your company's **Cost Model** deal with how you compensate your collaborators and resources for giving you the energy

---

[2]Opinion-shopping refers to the practice of dropping one auditor for another because the incumbent auditor has significant disagreements with company management on financial reporting policy, potentially compromising the ability of the company to get the unqualified opinion it is looking for.

you need to deliver happiness to your customers. Another way of putting it is they describe all of the costs you incur to operate your business, including operating leverage (the relative importance of fixed versus variable costs), economies of scale (cost advantages from growth in output), and economies of scope (cost advantages from a larger scope of operations).

- What are our organization's major costs?
- How do these costs compare, on a percentage basis, with our competitors? Can we explain these differences?
- How have cost percentages trended in recent years? What factors account for these changes?
- How much operating leverage (fixed vs. variable costs) do we have? Are efforts (e.g., outsourcing) underway to make more of our fixed-costs variable?
- What steps have we taken, or are we taking, to improve operating efficiency?
- What systems are in place to ensure continuous improvement in our key processes?
- What is the nature of our regulatory environment? What are the primary regulations (environmental, labor law, worker safety) that apply to our company? Through what means do we stay current on regulatory changes?
- Do we have the necessary compliance capabilities to deal with regulatory and legal changes when they happen?
- How do our tax rates compare to others in our industry? If our rates are higher or lower, why?
- How business-friendly is our home base?
- How efficient is the business infrastructure in our home base? If there are important inefficiencies, how do we work around them?

# Index